BIOINFORMATICS:
Databases and Systems

BIOINFORMATICS:
Databases and Systems

edited by

Stanley Letovsky

KLUWER ACADEMIC PUBLISHERS
Boston / Dordrecht / London

Distributors for North, Central and South America:
Kluwer Academic Publishers
101 Philip Drive
Assinippi Park
Norwell, Massachusetts 02061 USA
Telephone (781) 871-6600
Fax (781) 871-6528
E-Mail <kluwer@wkap.com>

Distributors for all other countries:
Kluwer Academic Publishers Group
Distribution Centre
Post Office Box 322
3300 AH Dordrecht, THE NETHERLANDS
Telephone 31 78 6392 392
Fax 31 78 6546 474
E-Mail <services@wkap.nl>

 Electronic Services <http://www.wkap.nl>

Library of Congress Cataloging-in-Publication Data

Bioinformatics: databases and systems / edited by Stanley Letovsky.
     p. cm.
  Includes bibliographical references.
  ISBN 0-7923-8573-X (alk. paper)
  1. Genetics--Data processing. I. Letovsky, Stanley.
QH441.2.B55 1999
576.5'0285--dc21
                                                 99-28476
                                               CIP

Printed on acid-free paper.

Printed in the United States of America

To Tish, Emily, Miles and Josephine

TABLE OF CONTENTS

SYSTEMS

INTRODUCTION

Stanley Letovsky

Bioinformatics is a field that has exploded into the awareness of biomedical, pharmaceutical and agricultural researchers in the last few years, in parallel with the equally new field of genomics. The central idea of genomics, first articulated by a few visionary biologists and funding officials at the US Department of Energy and National Institutes of Health in the 1980's, was to scale up the laboratory techniques of molecular biology, and to bring to biology a degree of completeness that seemed revolutionary at the time. Rather than mapping one gene at a time, people began to develop whole genome maps at specified marker densities. Rather than sequencing a gene at a time, they proposed to sequence the entire human genome, a project one hundred thousand times larger. Although the goal of a complete sequence of the 3000 megabase human genome remains in the future[1], the significance of this vision has already been demonstrated many times over. Major milestones to date include the first complete genomic sequencing in 1995 of the 2 MB genome of the bacterium *Haemophilus influenzae* [2] by a group at the Institute for Genomic Research (TIGR), followed by some dozen other bacterial species since then; of the first single-celled eukaryote in 1997, the budding yeast *Saccharomyces cereviseae,* with a genome size of 12 MB, by an international consortium of academic laboratories [3], and the completion in 1998 of the first multicellular eukaryote, the nematode worm *Caenorhabditis elegans*, with a genome size of about 100MB [4]. Projects are well under way to produce the genomic sequences of the fruitfly *Drosophila melanogaster* [5] and the mustard weed *Arabidopsis thaliana* [6], important genetic model organisms for the animal and plant kingdoms, respectively. Discussion is under way on a public project to sequence the mouse, which is a crucial model organism for the mammals.

In addition to genome sequencing, a number of other large-scale technologies have been developed to shed new light on the structure and function of organismal genomes. These include EST sequencing, which allows gene RNA products to be observed more directly than genomic sequencing permits; transcriptional profiling methods such as microchip-based hybridization arrays, that allow measurement of cells' gene expression levels; yeast 2-hybrid systems to allow the construction of protein interaction maps; radiation-hybrid mapping which allow

---

[1] The most recent estimate of the completion date for the public sector human sequencing project is the end of 2003 [1], with a "rough draft" now expected by mid-2000.

genome maps to be built (in some species) without reliance on cumbersome natural recombination; and high-throughput screening methods which allow the biological effects of large small-molecule compound libraries to be rapidly assessed. If genomics at present is using tomorrow's technologies today, often before all the kinks have been worked out, numerous groups are hard at work on the technologies of the day after tomorrow. Examples include protein profiling, or proteomics, which surveys the protein content of cells and tissues using high-resolution mass spectrometry; metabolic profiling, which measures the small molecule content of tissues; cheap polymorphism detection methods; and nanofabricated laboratory-on-a-chip technologies that may provide the elusive increases in speed and reductions in cost that have long been sought for "conventional" genomic technologies such as automated sequencing.

It is against the backdrop of this breakneck technology development and mass production of genomic data that the field of bioinformatics emerged. People had of course been applying computers to biological data for years before the term was coined, and most of the common algorithms for biological sequence comparison had been invented by 1980. But it was not until the mid-1990's that the field acquired a name, and suddenly became respectable, even fashionable. By 1996 it seemed that every other issue of *Science* contained an article bemoaning the desperate shortage of bioinformaticians in academic and industrial labs (e.g. [14]).

A crucial parallel development in the larger culture that coincided with the emergence of genomics and bioinformatics was the explosion of the Worldwide Web. Vindicating, perhaps, Marshall MacLuhan's cryptic insight that "the medium is the message" (or "mass age"), the Web has inserted itself into discipline after discipline, business after business, unleashing expanding ripples of transformation. Indeed the Web is one of the few technologies that are developing as fast as genomics, but the connection between them runs deeper than that. The Web turns out to be a nearly ideal vehicle for delivering genomic data to the scientific community; it is hard to imagine what bioinformatics would look like without it.

So what is bioinformatics? Definitions vary with the users of word; related terms like computational biology are held to be synonyms by some, and by others to reflect subtle distinctions. In practical terms there are some important distinctions to be made between the tasks of developing algorithms, of programming databases, and of curating database content. Computational biology algorithms for sequence comparison, sequence assembly, sequence classification, motif induction and recognition, and protein structure prediction have been the subject of several recent books [7-13] whereas the database system-building and content curation aspects have received less treatment[2]. These are perhaps the less glamorous aspects of the field, more lore than art, but these are also the areas where there seems to be a great hunger

---

[2] A noteworthy exception is the annual database issue of *Nucleic Acids Research* which each January allows public database providers the opportunity to report on recent developments in their projects. This book is intended to provide a complementary resource, with more freedom to explore particular aspects of the systems in depth.

for wisdom. Where algorithms tend to get implemented, packaged and shared as black boxes, systems get built over and over in different establishments, and people facing the same problems for the first time are always peppering their more experienced counterparts with questions like "did you use approach/product/standard X in your project? Was it any good? How do you keep the data up to date? How do you enforce content quality?" This book was conceived as a resource for people asking such questions, whose numbers seem to be doubling every six months at present. It is not possible to give definite answers to these sorts of questions – yes you should or no you shouldn't use ACEDB or CORBA or OPM or whatever. The software changes from month to month, the problems change, the options change. Today's technological rising star may be tomorrow's horror story, and vice versa. Even the insights derived from bitter experience can be questionable – did you really diagnose the problem correctly, and if you build the next system the way you now think you should have built the last one, will the result really be better or will you simply encounter a different, equally painful set of tradeoffs? Nonetheless experience is better than no experience, so I asked the contributors to this volume to include in their articles lessons learned from developing their systems, and to write down their thoughts on how they might do things differently -- or similarly -- if they were doing it again.

The contributors represent what I hope will be an interesting, albeit incomplete, sample of some of the most exciting work being done in bioinformatics today. The articles are somewhat arbitrarily divided into two sections: Databases and Software. The intent was that articles that focused more on content would fall into the former category, while articles that focused on technology would fall into the latter, but there were a number of borderline cases. My hope is that this collection will be of interest to readers who have arrived at the interdisciplinary world of bioinformatics either from the biology side or the computational side (as well as those more distant migrants from literature, history, business, etc.). The database articles may be more intelligible to readers with more of a biology background, and the software articles to readers with more software engineering; hopefully there is something to interest (and confuse) just about everyone.

The articles in the Database section represent some of the established (or in some cases recently disestablished!) citizens of the database world, as well some promising new efforts. The first few articles describe systems focused on the molecular level; these are mostly multispecies, comparative systems. Karl Sirotkin of NCBI describes some of the software underpinnings of Entrez, the most widely used molecular biology resource. The article on HOVERGEN by Duret et al describes an interesting new approach to integrating phylogenetic and coding sequence data into an organized whole. Several articles focus on the fast-developing area of metabolic and regulatory pathway databases, including those by Overbeek et al on WIT, Kanehisa on KEGG, and Karp and Riley on EcoCyc.

The remaining articles in this section describe primarily databases organized around organisms rather than molecules. Alan Scott describes the extensive literature-based curation process used to maintain the high quality standards of OMIM, the fundamental resource on human genetic disease. My own article looks at methods of integrating maps in the erstwhile human Genome Database (GDB). Cooper et al describe their database of human mutations, an early entry in the increasingly important field of variation databases.

The article by Nadkarni et al provides a link to the developing field of neuroinformatics, which is concerned with databases of such neuroscience data such as structural (MR or CT) or functional (PET, fMRI) images of the brain, histological slices, EEG and MEG data, cellular and network models, single cell recordings, and so on. [15]. This article is included not only as a representative of neuroinformatic work, but because it is one of the few current neuroinformatics efforts that links the molecular scale of bioinformatics to the neurophysiological scale, since it addresses the physiology of olfaction from the receptor sequences up to cellular and network physiology.

Eppig et al describe the Mouse Genome Database (MGD) and its companion system, the mouse Gene Expression Database (GXD). One of the key challenges for the next generation of databases is to begin to span the levels of organization between genotype and phenotype, where the processes of development and physiology reside. Baldock et al describe an anatomical atlas of the mouse suitable for representing spatiotemporal patterns of gene expression; the Edinburgh (Baldock et al) and Jackson Laboratory (Eppig et al) projects are collaborating to link the genetic and spatial databases together. The plant kingdom, which has recently experienced a rapid acceleration of genomic scrutiny in both the private and public sectors, is represented in articles on MaizeDB by Polacco and Coe and on the USDA's Agricultural Genome Information System by Beckstrom-Sternberg and Jamison. Gelbart et al describe the rich integration of genomic and phenotypic data on *Drosophila* in Flybase. Mary Berlyn describes the *E.coli* Genetic Stock Center Database, which provides query-by-genotype access to the stock center's extensive collection of mutant strains.

The Software section contains a number of articles that address one or another aspect of the problem of integrating data from heterogenous sources. There are two common ways to achieve such integration: *federation*, in which the data continue to reside in separate databases but a software layer makes them act as a single integrated collection, and physical integration, often called *warehousing*, in which the data are combined into a single repository for querying purposes. Both approaches involve transforming the data into a common format; federation does the transformation at query time, whereas warehousing does it as a preprocessing step. One consequence of this difference is that warehouses are more difficult to keep current as the underlying databases are updated. The choice of federation vs. warehousing has performance implications as well, though they are not always easy to predict. A warehouse can map in a straightforward way to a DBMS product, and make full use of the tuning and optimization capabilities of that product. Federated systems must pay the price of translating queries at run-time, possibly doing unoptimized distributed joins of query fragments across multiple databases, and converting data into the standard form. It is also possible for federated systems to gain performance by distributing queries across multiple processors in parallel, though such gains are rare in practice.

The OPM system, described in the article by Markowitz et al, uses a middleware layer to impose a uniform object-oriented data model[3] on a potentially heterogeneous set of back-end databases. Davidson et al's BioKleisli takes a similar approach but uses as its common data model a model adopted from logic programming which is similar to the relational model but more powerful.

The SRS system described by Carter et al occupies an interesting middle ground between federation and warehousing. In SRS the datasets are warehoused in a single computer for searching, but remain organized in their original formats. New databases are added to the warehouse by supplying parsers for their flat-file formats; the files themselves are unaffected, which makes updating an SRS system uniquely simple compared to most warehouse designs.

The Biology Workbench project exemplifies another approach to data integration made possible by the Web. It does not physically bring the data together like SRS, nor does it create a virtual unified database like BioKleisli or the OPM multidatabase capability. Instead, it integrates at the level of the front end by providing a thin veneer of user interface which provides access to a number of capabilities on the Web. A similar concept is employed by the BCM Search Launcher [16].

A key challenge in bioinformatics software development is the need to continuously evolve software systems to incorporate or adapt to new technologies while maintaining compatibility with existing (legacy) systems. Traditional software development practice has been described in terms of a "waterfall" model, in which development progresses continuously "downstream" from requirements analysis to design to implementation. This model provides little guidance to bioinformatics developers, whose task more closely resembles that of an auto mechanic trying to redesign a car while driving it. The rapid prototyping model, in which systems are built by successive approximation from crude first attempts, comes closer to the mark, but still assumes the luxury of an extended prototyping phase. The component-based design model, in which systems can be quickly assembled from reusable components, is one that many in bioinformatics are pinning their hopes on. Jungfer et al advocate the use of CORBA, an industry standard for designing distributed object-oriented software systems, which has been adopted at the European Bioinformatics Institute and elsewhere as a middleware layer to handle communication between back

---

[3] A *data model* is a set of primitive constructs, such as sets, relations or objects, which can be used to build database schemas. The most common data models are the relational and the object-oriented. A *data modelling language* expresses the concepts of some data model in a specific notation for writing down schemas; the SQL language used in relational databases includes a data modelling component called the create table statement. A *schema* describes the structure of data in a particular database. A *database management system* (DBMS) such as Oracle™ or Sybase™ interprets a schema definition written in a data modelling language as instructions for creating a database with the specified structure.

end databases and front-end viewers and analysis tools. In contrast to OPM and BioKleisli, CORBA does not so much offer a standard data model as provide a mechanism for isolating system components from the design constraints that might be imposed by choosing particular data models. The Biowidgets article by Crabtree et al describes an approach to modular software development for bioinformatics visualization. Biowidgets represents an attempt to apply that modular design philosophy to the problem of data visualization.

The ACE database manager, described by Jean and Danielle Thierry-Mieg, is the database kernel of the ACEDB system developed by Jean Thierry-Mieg and Richard Durbin for the *C.elegans* genomic database, and subsequently reused widely by many other projects. The tremendous success of ACEDB over the years can be attributed to a number of factors, including its biologist-friendly user interface, the ease with which data can be entered into it, the ease with which its schema can be modified, the ease with which new visualization and analysis routines can be added to it, the extensive genomics knowledge incorporated into its standard schema and software, and its price (free!). At the core of the system is a somewhat non-standard object-oriented DBMS which over a reasonably wide range of database sizes can outperform most commercial relational databases at tasks such as retrieving an object detail view, pulling up annotated sequence, or displaying a genetic map. It is heartening, in an era when DBMS development has been taken over by large corporations wielding enormous programming teams, to see that a tiny team of talented individuals can still write a system which outperforms the commercial products in many respects.

Laboratory information management systems (LIMS) are a species of software remote from the experience of many bioinformatics practitioners, for whom data are something that automatically appear in databases. Those of us who have had occasion to wander into the more upstream regions of the data production process encounter a world of robots, sequencing machines, bench biologists, pipettes and 96-well plates. This world turns out to be full of interesting and exotic software engineering challenges which are not well served by the current generation of commercial solutions. Goodman et al present an elegant and general approach to the problem of laboratory process workflow which they developed for use in the Whitehead Genome Sequencing Center.

The articles in this book constitute a very incomplete sample of the interesting systems out there. Notable gaps include databases for such important model organisms as yeast, C.elegans and Arabidopsis as well as a number of microbial databases; databases on protein sequences, families, and 3D structures; expression profile databases, and transcription factor and promoter databases. Starting points for finding such resources include [17,18].

One final note: in keeping with the rapid pace of change in this field, many of the authors are no longer at the institutions where the work was performed, and in some cases the addresses shown are out of date. During the period when this book was being written (1997-98), a number of the authors (myself, Markowitz et al, Karp, Etzold) moved from public sector to private sector positions, and several of the systems described (OPM, BioKleisli, EcoCyc, SRS, WIT, EBI-CORBA) went from research systems to commercial products.

I would like to express my thanks to the authors for their contributions, and to Anita Tilotta, Mary Panarelli and Cristina Carandang for their assistance in the preparation of this book.

References

1. F.S.Collins et al, New Goals for the U.S. Human Genome Project: 1998-2003. Science v.282, 23 Oct 1998, pp.682-689.
2. R.D.Fleischmann et al, *Whole-Genome Random Sequencing and Assembly of* Haemophilus influenzae *Rd*. Science v.269, 28 July 1995, p.496. See also http://www.tigr.org/tdb/mdb/hidb/hidb.html.
3. A. Goffeau et al, *The Yeast Genome Directory*, Nature 387 (suppl.) 1-105 (1997). See also http://genome-www.stanford.edu/Saccharomyces/.
4. The *C.elegans* Sequencing Consortium, *Genome Sequence of the Nematode C.elegans: A Platform for Investigating Biology*, Science v.282 11 Dec 1998 p.2012 and other articles in same issue. See also the *C.elegans* Project Web page http://www.sanger.ac.uk/Projects/C_elegans/.
5. The Berkeley Drosophila Genome Project, http://www.fruitfly.org/.
6. D.W.Meinke et al, Arabidopsis thaliana: A Model Plant for Genomic Analysis, Science v.282, 23 October 1998, p.662. See also Arabidopsis thaliana database: http://genome-www.stanford.edu/Arabidopsis/.
7. M.S.Waterman. *Introduction to Computational Biology: Maps, sequences and genomes*. Chapman & Hall 1995.
8. R.Durbin, S.Eddy, A.Krogh, G.Mitchison. *Biological Sequence Analysis: probabilistic models of proteins and nucleic acids.* Cambridge University Press (1998)
9. Steven Salzberg, David Searls, and Simon Kasif, editors. *Computational Methods in Molecular Biology*, Elsevier Science (1998).
10. P. Baldi and S. Brunak. *Bioinformatics: The Machine Learning Approach* MIT Press 1998.
11. A.D.Baxevanis and B.F.F.Ouellette, Eds. *Bioinformatics: A Practical Guide to the Analysis of Genes and Proteins.* Wiley-Interscience 1998.
12. M.J.Bishop, ed. *Guide to Human Genome Computing, 2nd edition* Academic Press 1998.
13. M.J. Bishop, editor: *Nucleic Acid and Protein Databases and How to Use Them,* Academic Press, in press 1998.
14. E. Marshall. *Bioinformatics: Hot Property: Biologists Who Co*mpute Science 1996 June 21; 272: 1730-1732.
15. The Human Brain Project Web Server, http://WWW-HBP.scripps.edu.
16. K.Worley. *BCM Search Launcher* http://kiwi.imgen.bcm.tmc.edu:8088/search-launcher/launcher.html.
17. WWW Virtual Library of Biomolecules, http://golgi.harvard.edu/sequences.html.
18. WWW Virtual Library of Biosciences, http://golgi.harvard.edu/biopages.html

DATABASES

1   NCBI: INTEGRATED DATA FOR MOLECULAR BIOLOGY RESEARCH

Karl Sirotkin

National Center for Biotechnology Information, National Library of Medicine, National Institutes of Health, Bethesda, MD 20894

Summary

Since 1992 the National Center for Biotechnology Information (NCBI) has provided integrated access to all public genetic sequence information and its associated annotation, as well as the citations and abstracts from the published literature referenced by the genetic sequence records. This chapter describes the main database that contains the genetic sequence data used by this integrated access, and how these data are linked both to other sequences and to the published literature. Entrez is the application used for accessing most of these data and their links and it can be used on a wide variety of hardware platforms. More recently, Web browser-based Entrez access has also been available (URL: http://www.ncbi.nlm.nih.gov/Entrez/). Subsets of these data are also available for Blast searching. It is hoped that this chapter will be a useful resource for both software developers and end users of this data.

Introduction

Interconnections between data help to integrate them. NCBI defines interconnections between genetic sequence data, structure data, taxonomic data, and literature references. These links may also be between the same type of records, for example, between literature articles. Articles are linked as similar using term content statistics [1,2,3]. Links between genetic sequence records are based on Blast sequence comparisons [4], linking similar, and thus possibly homologous sequences. Links between structure records are based on Vast structure comparisons [5], linking structures that tend to be similar.

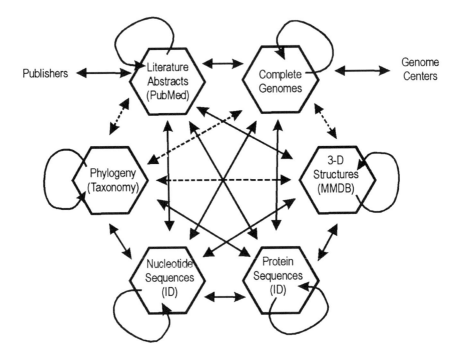

Figure 1: Data interconnected by NCBI.

Figure 1: Hexagons represent data that has been integrated, curved lines represent links between similar records, and straight lines between hexagons represent links between different data types, reciprocally linked, as described in the text. Dotted lines represent links under development.

Each record of one type can also link, reciprocally, to records of other types. Citations in the genetic sequence records determine links between the literature and genetic sequence information. Proteins translated from nucleic acid sequences, are linked to those sequences. Sequences can also be retrieved based on the phylogeny of the organism from which they come.

The data integrated by NCBI has recently been expanded. Now, most of the biological literature is available in a new database, PubMed, that includes all MEDLINE records plus some additional records. Links to electronic full-text articles have also been added where possible. Genomic and chromosomal maps linked to sequence information are available. A curated taxonomic tree is maintained [6] for organisms identified by the authors of the sequence submissions as the source of the genetic sequence data. This tree is independently available on the web (http://www.ncbi.nlm.nih.gov/ Taxonomy/taxonomy/home.html). Subsets of sequence records from organisms at taxonomic subtrees can be retrieved.

The rest of this chapter describes the management of the sequence data interconnected by NCBI. This data is represented by the two lower hexagons in figure 1, and all flows into the ID database, which is described in a subsequent section.

Conversion to ASN.1

Transforming data into the same format allows them to be used by software independently of the original source, a major requirement for integration. Within this uniform format the accuracy of the identifiers used to link, or point, to other records determines the accuracy and actual meaning of the links themselves. For literature articles, NCBI has added an integral identifier, the PubMed identifier. For convenience, many articles contain both a MEDLINE unique identifier and a PubMed identifier. Genetic sequence identifiers are more complex and can occur in different forms and places within each data record. Before this integration and use of sequence identifiers can be explained, the way in which NCBI represents the data must be clear.

So that the databases and other later applications can use a common format and be insulated from parsing the input data, the diverse streams of data are converted to Abstract Syntax Notation (ASN.1), an international standard [7]. When using ASN.1, one file describes the format to which other files (messages) of data must correspond. This file can be considered the definition of the message format. The data conforming to the message format can be understood by later application software that is thereby insulated from details of the original input formats. The messages use a "tag-value" format, in that an identifier describes the value that follows; however, analogous to programming language record definitions, the ability to use both recursion and user defined types in the definition of the message format allows for almost infinite complexity in the messages themselves. Since ASN.1 messages can thus become rather complex and are not intended to be human readable, other report formats, such as GenBank™ flatfile format, are used to display this data.

"asn.all" ASN.1 message definition

The particular format, or message definition, plays a central role. It is what describes the syntax into which all the sequence record information must be parsed. The original definition proposed by Jim Ostell [8] has been used with minor modifications for over five years. It is available as the file, "asn.all" in the NCBI toolkit distribution (ftp:ncbi.nlm.nih.gov; directory toolbox/ncbi_tools) and is discussed in detail in the NCBI Programmers Toolkit documentation [8]. "asn.all" describes the format of both the literature and genetic sequence messages. NCBI also makes use of other ASN.1 message definitions.

So that the data in the ASN.1 messages can be used by software, C language structures that map fairly closely to the "asn.all" definitions were designed, as well as software (object loaders) that could read and write these messages from C language structures to files and vice versa. These original structures and object loaders were hand crafted. More recently, the program "asncode" was written by the author and made available as part of the NCBI toolkit distribution. "asncode" takes ASN.1 message definitions and automatically generates both the C language structure definitions and object loaders. This "asncode"-generated software is used in network communication for the Entrez and Blast systems and the MMDB [9] structure manipulating software and could be easily used by software developers for most ASN.1 applications.

Sequence record types

In the "asn.all" definition, a "Bioseq" is a biological sequence that can contain information in addition to the sequence. For the purposes of this discussion, a Bioseq can be considered a collection of a "sequence" block, an "id" block, a "history" block, a "descriptor" block, and an "annotation" block. Each Bioseq can have a set of synonymous sequence identifiers that it contains in its "id" block. The semantics of the definition in "asn.all" are that this set of sequence identifiers are names for this Bioseq. Sequence identifiers that occur elsewhere in the record are pointers to Bioseqs that contain those sequence identifiers in their "id" block. "Descriptors" provide information that can apply to the entire sequence, such as citations to the literature and taxonomic classifications for the source of the material that led to the sequence information. Feature "annotation" applies to a particular region of the sequence, such as a coding region. These feature annotations use a sequence identifier to point to the Bioseq containing the sequence being described.

Bioseq sets

In the "asn.all" definition, a sequence entry can either be a Bioseq, or a more complex set of Bioseqs. One example of a more complex set is the set of protein Bioseqs combined with the nucleic acid that encodes them. The Bioseq can either contain actual sequence, or can incorporate sequence information solely by reference to other Bioseqs that actually contain the sequence. An example of this incorporation by reference is in the set of Bioseqs comprising the exons of a gene. This is the most common type of a "segmented set". In this case, the set begins with a Bioseq that points to the Bioseqs that contain the exon sequences. These pointers use sequence identifiers to specify the order of the exons by referencing the name (sequence identifier) of the Bioseqs containing the exon sequence. The Bioseqs containing the actual raw sequences for the exons are, in this case, part of the Bioseq set that includes the Bioseq pointing to them. Significantly, this "pointer" Bioseq, which has no sequence of its own and only incorporates sequence by reference, can be processed by the NCBI software system in the same way as any other Bioseq.

The entries in the Genomes division of Entrez are another example how pointers incorporate sequence data by reference to other Bioseqs. However, they differ from the sequence entry of the segmented set in that the Bioseqs containing the raw sequence are not in the same entry. This makes it critical for the sequence identifiers to be used accurately and uniquely.

There are other sets of data that can be grouped together. For example, sequences of the same gene from a group of related organisms comprise a population set, such as the actin sequences from related vertebrates. These and other more complicated sets are just beginning to be used and will not be discussed further in this chapter.

Sequence Identifiers.

There are a variety of reasons why there are multiple sequence identifiers in the "id" blocks of Bioseqs. For example, consider an expressed sequence tag (EST) sequence. In its "native" database, dbEST[10], it is given an integral tag, an "est_id" for its unique sequence identifier. If this sequence is to be used outside of the context of this particular database, the string "dbEST" must be added to this integral identifier for the resulting composite identifier to be unique. However, to appear in GenBank™[11], the sequence record needs, additionally, both a LOCUS name and an ACCESSION number. Every sequence in GenBank™ is also given a "gi" number, which allows all sequences to be retrieved with a single integral key, by Entrez, for example. Data from other sources also retain their identifiers assigned by those sources. So this single sequence record will have four to six different sequence identifiers. Different retrieval programs will use a different identifier from this synonymous set.

'gi's Vs accessions

There has been some confusion about the role of the 'gi' and how it compliments an ACCESSION. When a laboratory submits a sequence from a piece of DNA to a database, an ACCESSION number is assigned and is permanently associated with that piece of DNA. When the record containing that sequence is loaded into the ID database (see below) an initial 'gi' is assigned to that sequence. Further experiments over a time may alter the best understanding of the true sequence of that DNA. When these new sequences for the same piece of DNA are submitted to NCBI, a new 'gi' is assigned. This leads to a "chain" or series of sequences and corresponding 'gi's for the same piece of DNA. When it is important to identify the piece of DNA, for example as a subclone at a particular location within some other clone, then the ACCESSION is best used. When the particular sequence is most important, for statements about sequence similarity or some conceptual translation, the 'gi' that points to the particular sequence intended is best used. Statements and experiments that use 'gi's can always be repeated, because the sequence identified with a particular 'gi' is always available, even if the particular sequence identified by that 'gi' is not thought, currently, to be the accurate sequence. This relationship between ACCESSION and 'gi' is shown in Figure 2, below.

gi   gi   gi
100 — citation change → 100 — sequence change → 201

Figure 2: A "chain" of 'gi's

In this hypothetical example, the gi first assigned to this DNA sequence was gi=100. Although the citation in the sequence record was subsequently changed, the gi number was not. Only where the sequence information was altered was the gi changed to gi = 201. The number is arbitrary and reflects that more than a hundred sequences or sequence changes had been submitted in between the original submission of the gi=100 and the updated sequence (gi=201).

Not every sequence is given, currently, an ACCESSION number. In particular, protein sequences derived from conceptual translations are not assigned ACCESSIONS.

NCBI keeps a record of the history of 'gi's assigned to each sequence, even if that sequence does not have an ACCESSION. This information is present in the ASN.1 so that older records point to newer records and vice versa, making it possible to follow a trail of sequence changes for a particular piece of DNA (see below).

Integrating Database (ID)

As they are loaded into the ID database, which was designed and developed by the author, ASN.1 messages undergo uniform processing. This processing includes the assigning and tracking of 'gi' identifiers. All records for the GenBank™ updates and releases pass through ID.

During loading of a message into ID, the older record in the same chain is compared. Generally, since the messages are a complex set of Bioseqs, there will be an ACCESSION number, or related sequence identifier, in at least one of the Bioseqs, that can be used to find the older record. Sequences from Bioseqs that can be matched because they use identical sequence identifiers are compared. If the type of sequence changes (e.g., DNA to RNA) or the sequence itself changes, a new 'gi' is assigned, and the history record is added, both in the ID and in the ASN.1 message itself. If the 'gi' does not change, for example with a citation change, any history from the older record is copied into the incoming record to preserve the integrity of the 'gi' trail in the ASN.1 message.

Proteins from conceptual translations pose a special problem for this assignment and tracking, because of the present lack of ACCESSION numbers on protein Bioseqs. These protein Bioseqs are mostly part of the complex set of Bioseqs that includes the nucleic acid that encodes them. Five rules are applied in an attempt to match these protein Bioseqs between those in the current incoming record and those in the existing record. The first rule that is satisfied is used to declare a match. These rules are necessary because the incoming records only sometimes contain sequence identifiers on the protein Bioseqs that can be used to match Bioseqs between the old and the new record. Once matched, the sequences are compared in the same way as for the nucleic acid Bioseqs matched via accession for the purpose of 'gi' assignment. These rules are:

1. Explicit declaration by sequence identifier. For example, the 'gi' is sometimes in the ASN.1 message itself.

2. Explicit declaration by the history in the incoming record. Since NCBI accepts ASN.1 messages directly, the history slot can be used to explicitly declare the relationship between protein Bioseqs. Of course, these are only allowed if the nucleic acid Bioseqs of the complex sets containing the proteins are also properly related. Here, too, a 'gi' identifier or its equivalent can be used.

3. Matching by exact sequence identity. This actually happens quite frequently.

4. Matching by exact identity of location on the nucleic acid Bioseq. This can happen without sequence identity because of a prior error in translation, for example, caused by using an incorrect genetic code.

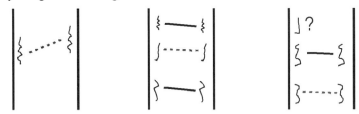

Figure 3: Deducing chains for protein 'gi's

Figure 3: Proteins (curly lines) Bioseqs in the new records (on the left) and old records (on the right) are ordered by their position on the nucleic acid (solid vertical lines). Proteins matched by one of the above rules are indicated by solid lines, while proteins matched by rule 5 (see the main text) are indicated by lighted dotted lines. A protein that can not be matched is indicated by the absence of lines and a question mark.

5. Matching by position in the set of records (Figure 3). Generally, the rule is if that a protein is bounded by either matched proteins or the end of the DNA sequence, it is assumed to be in the same chain (be encoded by the same gene). It is realized that this algorithm will occasionally make mistakes. However, it is usually correct and the trail of 'gi's can be very useful.

Conversion to 'gi' Sequence Identifiers

During loading into the ID database, sequence identifiers used as pointers to Bioseqs are converted to 'gi' type sequence identifiers. This allows any subpiece of the ASN.1 message to be used as an independent object by later software, even if separate from its original Bioseq.

A consequence of this conversion of all other pointer sequence identifiers to 'gi' identifiers is that if a record points to a sequence identifier not yet known to ID, the sequence identifier can not be converted. When such "sought" (currently unknown) identifiers are defined by having their sequence loaded into ID, the original record is altered to point to it. This provides some independence from the order of addition of data to ID and guarantees that all the sequence identifiers that can be converted to 'gi' identifiers have been converted, but with computational cost and increased complexity in the processing code.

Frequently, the information in one record "takes over" the information in others. For example, an author submits a new large sequence record that incorporates information from other smaller records. When desired, this new record receives its own new ACCESSION. In the current ID database and in the ASN.1 message, no distinction is made between this takeover and simply a new sequence for the same piece of DNA.

Scale-up for the Expected Flood of Genome Project Data

The ID database has fulfilled its role since early 1993. However, to allow scale-up and massive throughput, ID has recently been redesigned and a new system is under development. This system will distribute many of the tasks and storage that are presently centralized and instead will keep only limited information centrally. Specifically, a number of "satellite" databases, internal to NCBI, will hold the data themselves and will communicate with the main ID database, to be called IdMain, to receive 'gi' identifiers and to inform IdMain of the multiple sequence identifiers that map to this 'gi'. Communication is planned to be either synchronous through a direct connection, or asynchronous, using the Sybase Replication Server [12]. Because of the less tight coupling between the data and IdMain than is in the current ID, every Bioseq needs a primary sequence identifier to serve the same role as an ACCESSION and thus stays the same across different submissions of the same Bioseq.

The plan is to have access to the ASN.1 messages through a Sybase Open Server [12], which will be able to access information from IdMain to find which satellite database has the information needed to generate the ASN.1 message, or the ASN.1 message itself. This has already been implemented (Eugene Yaschenko, personal communication), using the original ID database as the data source. When the new distributed system is ready, there will be no changes to Entrez, or other applications retrieving data this way, just upgrades to the Sybase Open Server.

This will allow, for example, dbEST to hold and deliver EST sequences, without them having to be loaded into an additional database. Not only does this save computational resources and space, but it allows the greatest possible advantage to be made of the relational nature of dbEST itself. Furthermore, by allowing for increased volume by the addition of multiple satellite databases that hold ASN.1 messages and are either relational or non-relational, scale-up should be efficient.

Acknowledgments

The authors wishes to thank to Jo McEntyre for helpful comments on the manuscript.

References

1. G. Salton, "Automatic Text Processing," Addison-Wesley, Reading, Massachusetts. 1989.

2. W. J. Wilbur, **A retrieval system based on automatic relevance weighting of search teams**, Proceedings of the 55th American Society of Information Science, Annual meeting (Ed.) D. Sahr (1992(, Pittsburg, PA. 29, 216-220.

3. W. J. Wilbur and Y. Yang, **An analysis of statistical term strength and it use in the indexing and retrieval of molecular biology texts**, Computers and Biology in Medicine, 26, 209-222.

4. S. F. Altschul, W. Gish, W. Miller, E. W. Myers, and D. J. Lipman, **Basic local alignment search tool**, Journal of Molecular Biology, 215, 403-410 (1990).

5. J. F. Gibrat, T. Madej, S. H. Bryant, **Surprising similarities in structure comparison**, Current Opinion in Structural Biology, 6, 377-385 (1996).

6. Improving GenBank's Taxonomy, NCBI News February, 1994.

7. M. T. Rose, "The Open Book: A Practical Perspective on OSI", Prentise Hall, Englewood Cliffs, New Jersey, 1990.

8. NCBI Software Toolkit Manual (1993) National Center of Biotechnology Information, Bldg. 38A, NIH, 8600 Rockville Pike, Bethesda MD 20894.

9. T. Madej, J.-F. Gibrat, and S. H. Bryant, **Threading a database of protein cores**, Proteins, 23,356-369 (1995)

10. M. S. Boguski, T. M. Lowe, and C. M. Tolstoshev, **dbEST - database for 'expressed sequence tags**, Nature Genetics 4, 332-333 (1993)

11. D. A. Benson, M. S. Boguski, D. J. Lipman, and J. Ostell, **GenBank**, Nucleic Acids Research, 25, 1-6 (1997)

12. For information on Sybase products, see their web pages, Http://sybooks.sybase.com/cgi-bin/nph-dynaweb

2 HOVERGEN: COMPARATIVE ANALYSIS OF HOMOLOGOUS VERTEBRATE GENES

Laurent Duret, Guy Perrière and Manolo Gouy

Laboratoire de Biométrie, Génétique et Biologie des Populations, UMR CNRS 5558, Université Claude Bernard, 43 Bd du 11 Novembre 1918, 69622 Villeurbanne cedex, France

Introduction

Comparison of homologous sequences is an essential step for many studies related to molecular biology and evolution: to predict the function of a new gene, to identify important regions in genomic sequences, to study evolution at the molecular level or to determine the phylogeny of species. The importance of comparative sequence analysis (comparative genomics) for deciphering the genetic information embedded in the human genome is now widely recognized, and thus, projects have been set up to sequence large genomic regions of model vertebrate organisms such as mouse [1] or pufferfish [2].

Databases already contain a considerable amount of vertebrate data suitable for comparative analysis. For example, more than 1,200 homologous genes between man and rodents are available in databases [3]. Besides mammals, large sets of sequence data are also available for birds (chicken), amphibians (xenopus) and bony fishes (zebrafish, pufferfish, salmonidae), thus covering a wide range of evolutionary distances, from 80 to 450 million years of divergence.

Thanks to this large and rapidly increasing amount of homologous sequences, comparative sequence analysis should be very efficient for improving our knowledge of the structure, function and evolution of vertebrate genomes. However, the search for homologous genes and interpretation of homology relationships are complex tasks that require to simultaneously handle multiple alignments, phylogenetic trees, taxonomic data and sequence-related information. Genomic sequence databases such as GenBank [4] or EMBL [5] do not include information relative to homology relationships between genes, and hence these analyses have to be done manually, which is very tedious and error prone.

To respond to these problems, we have developed a database of homologous vertebrate genes named HOVERGEN [6]. This database integrates protein multiple alignments, phylogenetic trees, taxonomic data, nucleic and protein sequences, and GenBank sequence annotations. With its graphical interface, HOVERGEN allows one to rapidly and easily select sets of homologous genes and evaluate homology relationships between sequences. This chapter describes the content of this database, the procedure we use to maintain it, a few examples of application and the future developments that we plan.

Definitions: homology, orthology, paralogy

Two sequences are said to be homologous if, and only if, they share a common ancestor [7]. In practice, homology is generally inferred from sequence similarity. Sequence similarity is not always a proof of homology: when the similarity is low, covering a short region, it is possible that it is due to structural or functional convergence or simply to chance [8]. In some cases, sequence similarity is only due to compositional biases (low complexity segments, such as proline- or alanine-rich regions) [9]. In the absence of compositional bias, when protein sequence similarity is at least 30% identity over 100 residues or more, it is almost certain that sequences share a common ancestor [10]. This definition of homology also explains why statements such as "these two proteins are 35% homologous" are incorrect although they are very frequently used: homology is an all or none property which can only be true or false.

In some cases, sequences are not homologous over their entire length: some proteins are constituted of modules (or domains) that have different evolutionary origins [11,12] These proteins are said to be mosaic (or modular).

Among sequences that are homologous over their entire length, one has to distinguish orthologous sequences, *i.e.* sequences that have diverged after a speciation event, from paralogous sequences, *i.e.* sequences that have diverged after duplication of an ancestral gene (Figure 1) [7]. This distinction is essential for molecular phylogeny since it is necessary to work with orthologous genes to infer species phylogeny from genes phylogeny. This distinction is also important to predict the function of a new gene or to search for functionally conserved regions by comparative analysis, because paralogous genes, even if closely related, may have different functions or regulations.

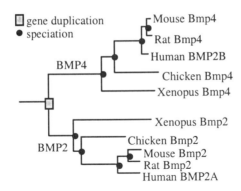

Figure 1: Phylogenetic tree of the BMP gene family illustrating paralogy and orthology concepts. BMP2 sequences from rat, mouse, man, chicken and xenopus are orthologous since they all diverged after speciation events. BMP2 and BMP4 genes are paralogous since they result from a gene duplication. As often, the gene nomenclature using 2A/2B for human genes and 2/4 for others does not clearly reflect the evolutionary history of these genes.

The distinction between orthologous and paralogous genes requires a careful analysis of homologous sequences [6]. First of all, it is not possible to rely on database definitions to identify orthologous genes, because many sequences are not annotated, and when annotations are present they are sometimes inexplicit or inaccurate. Thus, some paralogous genes may have very similar definitions, whereas orthologous genes may be given totally different names. Hence, it is necessary to use sequence similarity criteria to search for homologous genes.

Similarity search programs such as BLAST [13] or FASTA [14] allow one to quickly search for homologues in sequence databases. However, pairwise sequence comparisons are not sufficient to distinguish orthologous from paralogous genes. For example, suppose that one compares human bone morphogenetic protein-2B (BMP2B, GenBank accession number M22490) to mouse bone morphogenetic protein 2 (Bmp2, L25602). These two sequences are highly similar, they have similar definitions, and thus they might be considered as orthologous. However, the phylogenetic tree of all available BMP2B homologues clearly shows that mouse Bmp2 sequence is much more closely related to xenopus or chicken Bmp2 genes than to human BMP2B (Figure 1). Since mouse is much more closely related to man than to birds or amphibians, it is likely the BMP2B and Bmp2 genes are paralogous. Indeed, the phylogenetic tree shows that human BMP2B gene is orthologous to mouse Bmp4, and that mouse Bmp2 gene is orthologous to human BMP2A. It is important to notice that in the above example, in the absence of the other BMP genes from different species it would not have been possible to detect paralogy.

This example shows that it is necessary to compare all available homologous sequences to be able to detect paralogy. Thus, as far as genomes will not have been completely sequenced, it will not be possible to definitively demonstrate orthology between sequences. Rather, orthology relationships have to be re-analyzed each time a new gene is sequenced.

HOVERGEN: scope and content

HOVERGEN is a database dedicated to evolutionary or comparative studies. Its main goal is to allow one to quickly retrieve sets of orthologous genes (or proteins) among vertebrate species.

HOVERGEN contains all vertebrate sequences (genomic or RNA) available in GenBank (along with the corresponding annotations), except EST sequences. We decided in 1995 to exclude ESTs from HOVERGEN because most of them are partial, contain sequencing errors, are not annotated, and thus are of little value for our purpose. Initially, HOVERGEN contained only sequences from the nuclear genome [6], but now mitochondrial sequences are also included. Protein-coding sequences (CDS) can be automatically translated according to sequence annotations. Thus, HOVERGEN can be used both as a nucleic acid or a protein-sequence database.

The main information that is included in HOVERGEN is the entire classification of all protein-coding sequences into gene families:

- Coding-sequences are all compared between each others (at the protein level).

- Sequences that are homologous over their entire length are classified into families.

- Protein multiple alignments are computed for each family.

- Genes phylogenetic trees are derived from protein multiple alignments.

Furthermore, we also include in HOVERGEN information relative to genes structure that is not provided by GenBank, such as a precise description of non-coding sequences, location of CpG islands, and indication of GC-content (see [6]). Finally, to limit the problem of redundancy in sequence databases, protein-coding sequences (CDS) are all compared between each others to identify multiple sequences of a same gene.

The HOVERGEN database is not intended to study the modular evolution of proteins. Thus homology relationships between module or domains are not described in HOVERGEN.

Treatment of data

Identification of redundancy

It is common to find in DNA databases several entries that correspond to a same gene. For example, one entry may describe the mRNA and another the genomic fragment. In some cases, a same gene has been independently sequenced by different groups, or has been sequenced several time from different individuals to study polymorphism. In principle, one should find in databases only one entry for each gene, and if it is polymorphic, then allelic variations should be described in the annotations. But in practice, all redundant sequences are entered in databases, and there is no merging of partial overlapping sequences.

This redundancy is very problematic, not only because it gives a confuse view of the status of these redundant sequences (are they identical? splicing or allelic variant of a same gene? paralogous genes?), but also because it can considerably bias the results of statistical analyses.

In HOVERGEN, we systematically compare all CDSs between each other (with BLASTN [13]) to try to identify those that correspond to a same gene. As previously discussed [6], the problem with that approach is that two redundant CDSs may show some differences due to polymorphism, sequencing errors, or annotation errors. Taking into account published estimates of sequence polymorphism [15], and sequencing error rates [16-18], we decided to consider as redundant all CDS that share more than 99% identity (at the DNA level), and have less than 3 bases of difference in length. Using these criteria, we detected 21% of redundancy among the 63,520 vertebrate CDSs available in GenBank (release 101, June 1997). This level of redundancy is remarkably high, and it is thus necessary to take it into account when doing statistical analyses on sequence databases.

Redundancy is not eliminated from HOVERGEN because each entry may be associated to useful information. Rather, redundancy is explicitly declared, using a new qualifier ('redundancy_ref') that is included in sequence annotations. This qualifier is unique for each set of redundant CDSs. Thus, this information can easily be used to eliminate redundancy when required.

It is important to note that two homologous genes resulting from a recent duplication, speciation or conversion may be more than 99% identical (*e.g.* human α-1 and α-2 globin genes). Thus, declaration of redundancy in HOVERGEN should not be taken into account when one wants to study recent evolutionary events (< 4 million years) [6].

Classification of sequences into families of homologous genes

Sequence selection and similarity search. All available CDSs are classified, except partial CDSs shorter than 300 nt (about 25% of all CDSs). When several redundant CDS are available, only one is analyzed for the classification. CDSs are translated into proteins and compared between each others with BLASTP [13], using the PAM 120 score matrix. The threshold score to report similarity (S parameter in BLASTP) is set according to proteins length (L): S=150 for L • 170 aa, S=L-20 for L<170 and S=35 for L<55 aa. This threshold is high enough to avoid excessive noise due to low complexity segments present in many proteins. It is low enough to detect relatively distant similarities (• 30% identity over 150 aa or more). It should be noted that, as already discussed, this search is not intended to detect distant similarities between short protein modules.

Classification. The aim of the classification is to group together all sequences that are homologous over their entire length, without making distinction between orthologous and paralogous genes. However, when gene families are very large (*e.g.* there are more than 250 vertebrate globin-related sequences in HOVERGEN), computing of multiple alignments and graphical display of phylogenetic trees become very problematic. Thus, in such cases, clearly paralogous genes (*e.g.* α and β globin genes) are classified into distinct subfamilies.

The classification is done in two steps. A first programs groups all proteins sharing similarities. Note that at this step, if A is similar to B, and B to C, then A, B and C are grouped together, even if A and C share no similarity. Then, within each group, sequences are sorted according to similarity criteria with a program based on the UPGMA algorithm [19]. The distance metrics used for this process derives from the Poisson probability computed by BLASTP. Once groups are sorted, a manual expertise is required to split these groups into families of sequences homologous over their entire length.

Currently, 39,797 non-redundant CDSs are classified into 5,541 families (HOVERGEN release 25, July 1997). Among those CDSs, 2,232 (6%) have no homologue in their family. Figure 2 gives the distribution of families according to the number of sequences they contain (redundancy excluded).

Among the largest gene families, one essentially finds genes involved in the immune system, that are heavily studied for their polymorphism (Table 1). Most of families that are represented in a large number of different species correspond to mitochondrial genes, that are often used for phylogenetic studies (Table 2).

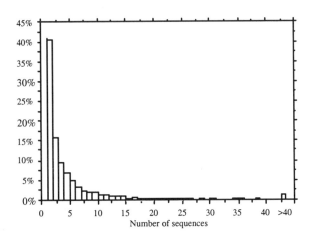

Figure 2: Distribution of families according to the number of sequences they contain (without redundancy) [HOVERGEN release 25, July 1997].

Table 1: The 10 largest families, redundancy excluded	
Immunoglobulins	6079 CDSs
Mitochondrion cytochrome b	1679 CDSs
MHC class I	636 CDSs
T-cell receptor beta-chain	564 CDSs
T-cell receptor alpha-chain	527 CDSs
MHC class II	452 CDSs
Collagen	283 CDSs
Cytochrome P-450	190 CDSs
Zinc finger protein	181 CDSs
Beta-B globin	162 CDSs

Table 2: Families represented in a large number of species	
Mitochondrion cytochrome b	1334 species
Mitochondrion cytochrome c oxydase subunit II COX-2	116 species
Mitochondrial NADH dehydrogenase subunit 4 NADH-4	90 species
Mitochondrion cytochrome c oxydase subunit I COX-1	90 species
Mitochondrial NADH dehydrogenase subunit 2 NADH-2	86 species
Beta-B globin	76 species
Immunoglobulins	61 species
Growth hormone	61 species
MHC class I	58 species
Protamine 1	58 species

Protein multiple alignment and phylogenetic trees

Protein multiple alignments are computed with CLUSTALV [20] for each gene family, except those that contain more than 150 sequences (Table 1). When several redundant CDSs are available, only one is included in the alignment. Phylogenetic trees are inferred from multiple alignments using the 'neighbor joining' method [21].

Updating of data and rate of growth

HOVERGEN is updated every two GenBank releases (every four months). New or modified sequences from GenBank are compared to HOVERGEN sequences and between each others, first at the DNA level to identify redundancy, and then at the protein level to update the classification. Multiple alignments and phylogenetic trees are computed for all families that have been modified or newly created. The rate of growth of the database (total amount of sequences) is exponential-like, with a doubling time of less than two years, but the number of gene families evolves rather linearly with time (Figure 3).

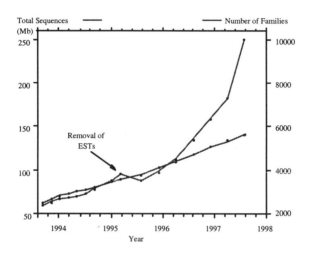

Figure 3: Evolution of the size of the HOVERGEN database

Database query software

Sequence retrieval

Sequences and annotations are managed using the ACNUC system [22]. This system is associated with a graphical retrieval software QUERY_WIN, that allows the user to build complex queries using multiple criteria [23]. Notably, thanks to the information that we include in HOVERGEN, it is possible to quickly select all homologous sequences available in a given set of species.

Graphical interface for multiple alignments and phylogenetic trees

We have developed a graphical interface that allows one to simultaneously handle all the data available in HOVERGEN to analyze homology relationships: phylogenetic trees, multiple alignments, taxonomic information and sequence annotations (Figure 4). The core of the interface is a phylogenetic tree viewer. Genes are colored according to the species from which they have been sequenced. The color that are affected to each taxa are either automatically set for different taxonomic level, or can be manually chosen by the user. When the user clicks on a gene in the tree, the sequence definition is displayed in another window. If there are several redundant sequences for that gene, all definitions are shown. Then, by clicking on a definition the user can visualize all the annotations associated to these sequences. Similarly, one can select from the tree the genes to be displayed in the multiple alignment. Alignments between selected genes are not computed *ab initio*, but reconstructed from the whole family multiple alignment (which is almost instantaneous).

WWW server

We have developed a WWW interface that allows one to query HOVERGEN and to download multiple alignments and phylogenetic trees through INTERNET (http://acnuc.univ-lyon1.fr/start.html) [24]. We also provide dedicated helper applications to directly visualize alignments and phylogenetic trees from the Web browser application. Although this server does not provide all the facilities of the graphical interface, it is useful for those who cannot install HOVERGEN on their local computer.

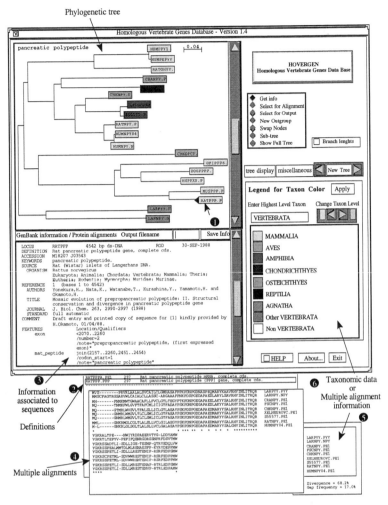

Figure 4: HOVERGEN graphical interface. When the user clicks on a gene in the tree (1), definitions of all corresponding sequences are displayed (2). By clicking on a definition, the user can visualize the annotations associated to the sequence (3). The same window is used to display protein alignments (4). Sequence divergence and indel frequency are indicated (5). Window (6) indicates the colors corresponding to the different taxa.

Phylogenetic tree

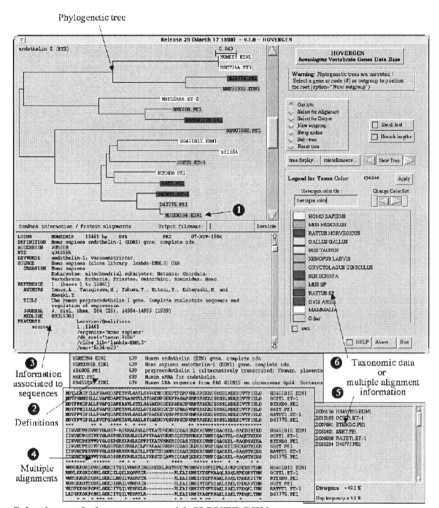

Selecting orthologous genes with HOVERGEN

As explained in the introduction, the search for orthologous genes is a complex task that requires a careful analyses of all available data. With HOVERGEN, this search can be done in two steps. The first step consists in using the QUERY_WIN sequence retrieval software to search for homologous genes known in different species, according to the criteria defined by the user. Then, the user has to examine with the graphical interface all the gene families to which these genes belong to distinguish orthologous from paralogous genes.

Examples of application

HOVERGEN has been used in various studies based on statistical analysis of orthologous genes, to study the evolutionary rate of genes [3,25,26], the structure of vertebrate genomes [27-29] or the statistics of genes size [30,31]. In molecular phylogeny, it is important to work on large samples of genes, and it is absolutely necessary to exclude paralogues from the analyses to be able to resolve phylogenetic trees of species. Thanks to the amount of data available in HOVERGEN it has been possible to address several important questions on vertebrates phylogeny, and some unexpected results regarding mammals or birds phylogeny have come out [32-34].

HOVERGEN is also useful as a tool to display a global view of all data available for a given gene family. Thanks to the graphical interface, the user can quickly access to the annotations of the various redundant sequences of a gene, or of the homologues of that gene in the same genome or in other species. Thus, HOVERGEN allows one to enrich the knowledge on a given gene by the information available for its various homologues.

Finally, HOVERGEN is also a powerful tool for correcting annotation errors. Newly sequenced genes are now often annotated simply by similarity with an already known sequence. Unfortunately, many errors occur because paralogous genes are not distinguished from orthologous genes. Such errors can easily be checked by looking at the phylogenetic trees provided by HOVERGEN.

Comparison with other databases

Several other databases include information on homology relationships between sequences. For example, ENTREZ [35] shows all significant similarities detected between protein or DNA sequences. However, ENTREZ does not provide multiple alignments nor phylogenetic trees. The PIR database now provides an entire classification of all published protein sequences, along with multiple alignments [36]. However, contrarily to HOVERGEN, PIR is limited to proteins, and does not give access to the gene sequences. Currently, HOVERGEN is the only database that provides an entire classification of vertebrate genes, with both multiple alignment and phylogenetic trees. Its main originality comes from its query retrieval system that allows one to quickly retrieve sets of homologous genes from different species and its graphical interface that allows one to analyze their evolutionary relationships.

Limitations

Several limitations of our database have to be mentioned. First of all, the graphical display of phylogenetic trees is limited to families having less than 150 sequences. Currently, only 10 families contain more than 150 sequences (Table 1), but this number should rapidly increase. Due to this problem of displaying large trees, we sometimes split paralogous genes from a same families into different subfamilies.

Thus, HOVERGEN allows one to do exhaustive searches of orthologous genes, but not of paralogous genes.

It should be noted that alignments provided by HOVERGEN are crude results from CLUSTALV, without any manual correction. Moreover, when calculating phylogenetic trees, one should exclude ambiguous parts of the alignment, which cannot be done in our automatic procedure. We also have noted some problems in phylogenetic trees (such as negative branches) that are due to the fact that we include partial sequences in our classification. Thus, although trees provided by HOVERGEN are very useful for analyzing evolutionary relationships, one should not take them as exact phylogenetic trees.

Perspectives

Up to now, HOVERGEN was limited to vertebrates, but we plan to extend this database to all organisms. The classification will be exhaustive, so that it will be possible to systematically analyze not only orthologues but also paralogues. This new database will be based both on GenBank/EMBL (for DNA sequences), and on SWISSPROT-TREMBL [37] to have access to high quality annotations for protein sequences. The graphical interface of this system is being written in Java, so that it will be possible to use it from any computer system for which the Java Virtual Machine is available. Also, as the Java libraries include tools allowing simple INTERNET connections and client/server interactions, it will be possible to query this database either locally or remotely.

Availability

HOVERGEN is available by anonymous FTP from our server (ftp://biom3.univ-lyon1.fr/pub/hovergen/) or from the NCBI (ftp://ncbi.nlm.nih.gov/repository/hovergen/). The QUERY_WIN program and HOVERGEN graphical interface are available for UNIX platforms (Sun, SGI, DEC-Alpha, IBM-RS/6000). HOVERGEN can be used through the Web at: http://acnuc.univ-lyon1.fr/start.html/. It is also possible to ask for an account at the UK MRC HGMP Resource Centre (http://www.hgmp.mrc.ac.uk/) or at the French INFOBIOGEN server (http://www.infobiogen.fr/) to use HOVERGEN remotely with a X-window emulating software.

Acknowledgments

This work is supported by the French Centre National de la Recherche Scientifique (CNRS).

34

References

1. Hood, L., Koop, B., Goverman, J. Hunkapiller, T. *Model genomes: the benefits of analysing homologous human and mouse sequences*, Trends Biotechnol., 1992, **10**, pp.19-22.
2. Brenner, S., Elgar, G., Sandford, R., MacRae, A., Venkatesh, B. Aparicio, S. *Characterization of the pufferfish (Fugu) genome as a compact model vertebrate genome*, Nature, 1993, **366**, pp.265-268.
3. Makalowski, W., Zhang, J.H. Boguski, M.S. *Comparative analysis of 1196 orthologous mouse and human full-length mrna and protein sequences*, Genome Res., 1996, **6**, pp.846-857.
4. Benson, D.A., Boguski, M.S., Lipman, D.J. Ostell, J. *GenBank*, Nucleic Acids Res., 1997, **25**, pp.1-6.
5. Stoesser, G., Sterk, P., Tuli, M.A., Stoehr, P.J. Cameron, G.N. *The EMBL Nucleotide Sequence Database*, Nucleic Acids Res., 1997, **25**, pp.7-13.
6. Duret, L., Mouchiroud, D. Gouy, M. *HOVERGEN: a database of homologous vertebrate genes*, Nucleic Acids Res., 1994, **22**, pp.2360-2365.
7. Fitch, W.M. *Distinguishing homologous from analogous proteins*, Syst. Zool., 1970, **19**, pp.99-113.
8. Doolittle, R.F. *Convergent evolution: the need to be explicit*, Trends Biochem. Sci., 1994, **19**, pp.15-18.
9. Wootton, J.C. Federhen, S. *Statistics of local complexity in amino acid sequences and sequence databases*, Computers Chem., 1993, **17**, pp.149-163.
10. Doolittle, R.F. *Searching through sequence databases*, Methods Enzymol., 1990, **183**, pp.99-110.
11. Patthy, L. *Modular exchange principles in proteins*, Curr. Opin. Struct. Biol., 1991, **1**, pp.351-361.
12. Patthy, L. *Introns and exons*, Curr. Opin. Struct. Biol., 1994, **4**, pp.383-392.
13. Altschul, S.F., Gish, W., Miller, W., Myers, E.W. Lipman, D.J. *Basic local alignment search tool*, J. Mol. Biol., 1990, **215**, pp.403-410.
14. Pearson, W.R. Lipman, D.J. *Improved tools for biological sequence comparison*, Proc. Natl. Acad. Sci. USA, 1988, **85**, pp.2444-2448.
15. Li, W.H. Sadler, A. *Low nucleotide diversity in man*, Genetics, 1991, **129**, pp.513-523.
16. Krawetz, S.A. *Sequence errors in GenBank: a means to determine the accuracy of DNA sequence interpretation*, Nucleic Acids Res., 1989, **17**, pp.3951-3957.
17. Kristensen, T., Lopez, R. Prydz, H. *An estimate of the sequencing error frequency in the DNA sequence databases*, DNA Seq., 1992, **2**, pp.343-346.
18. Lamperti, E.D., Kittelberger, J.M., Smith, T.F. Villa-Komaroff, L. *Corruption of genomic databases with anomalous sequence*, Nucleic Acids Res., 1992, **20**, pp.2741-2747.
19. Sokal, R.R. Michener, C.D. *A statistical method for evaluating systematic relationships*, Univ. Kansas Sci. Bull., 1958, **28**, pp.1409-1438.
20. Higgins, D.G., Bleasby, A.J. Fuchs, R. *CLUSTAL V: improved software for multiple sequence alignment*, Comp. Appl. Biosci., 1992, **8**, pp.189-191.

21. Saitou, N. Nei, M. *The neighbor-joining method: a new method for reconstructing phylogenetic trees*, Mol Biol Evol, 1987, **4**, pp.406-425.
22. Gouy, M., Gautier, C., Attimonelli, M., Lanave, C. Di-Paola, G. *ACNUC- a portable retrieval system for nucleic acid sequence databases: logical and physical designs and usage*, Comp. Appl. Biosci., 1985, **1**, pp.167-172.
23. Perriere, G., Gouy, M. Gojobori, T. *NRSub: a non-redundant data base for the Bacillus subtilis genome*, Nucleic Acids Res., 1994, **22**, pp.5525-5529.
24. Perriere, G. Gouy, M. *WWW-Query: An on-line retrieval system for biological sequence banks*, Biochimie, 1996, **78**, pp.364-369.
25. Mouchiroud, D., Gautier, C. Bernardi, G. *Frequencies of synonymous substitution in mammals are gene-specific and correlated with frequencies of non-synonymous substitutions*, J. Mol. Evol., 1995, **40**, pp.107-113.
26. Hughes, A.L. Yeager, M. *Comparative evolutionary rates of introns and exons in murine rodents*, J. Mol. Evol., 1997, **45**, pp.125-130.
27. Caccio, S., Zoubak, S., D'Onofrio, G. Bernardi, G. *Nonrandom frequency patterns of synonymous substitutions in homologous mammalian genes*, J. Mol. Evol., 1995, **40**, pp.280-292.
28. Bernardi, G., Hughes, S. Mouchiroud, D. *The major compositional transitions in the vertebrate genome*, J. Mol. Evol., 1997, **44**, pp.S44-S51.
29. Robinson, M., Gautier, C. Mouchiroud, D. *Evolution of isochores in rodents*, Mol. Biol. Evol., 1997, **14**, pp.823-828.
30. Duret, L., Mouchiroud, D. Gautier, C. *Statistical analysis of vertebrate sequences reveals that long genes are scarce in GC-rich isochores*, J. Mol. Evol., 1995, **40**, pp.308-317.
31. Ogata, H., Fujibuchi, W. Kanehisa, M. *The size differences among mammalian introns are due to the accumulation of small deletions*, FEBS Lett., 1996, **390**, pp.99-103.
32. Graur, D., Duret, L. Gouy, M. *Phylogenetic position of the order Lagomorpha (rabbits, hares and allies)*, Nature, 1996, **379**, pp.333-335.
33. Graur, D., Gouy, M. Duret, L. *Evolutionary affinities of the order Perissodactyla and the phylogenetic status of the superordinal taxa Ungulata and Altungulata*, Mol. Phylogenet. Evol., 1997, **7**, pp.195-200.
34. Hedges, S.B., Parker, P.H., Sibley, C.G. Kumar, S. *Continental breakup and the ordinal diversification of birds and mammals*, Nature, 1996, **381**, pp.226-229.
35. Schuler, G.D., Epstein, J.A., Ohkawa, H. Kans, J.A. *Entrez: molecular biology database and retrieval system*, Methods Enzymol., 1996, **266**, pp.141-162.
36. Mewes, H.W., Albermann, K., Heumann, K., Liebl, S. Pfeiffer, F. *MIPS: A database for protein sequences, homology data and yeast genome information*, Nucleic Acids Res., 1997, **25**, pp.28-30.
37. Bairoch, A. Apweiler, R. *The SWISS-PROT protein sequence data bank and its supplement TrEMBL*, Nucleic Acids Res., 1997, **25**, pp.31-36.

3 WIT/WIT2: METABOLIC RECONSTRUCTION SYSTEMS

Ross Overbeek, Niels Larsen, Natalia Maltsev, Gordon D. Pusch, and Evgeni Selkov

Argonne National Laboratory, Argonne, IL 60439

Introduction: What Is Metabolic Reconstruction?

For the past few years, we have been developing metabolic reconstructions for organisms that have been sequenced, and we have made a number of these working models available. By the term *metabolic reconstruction* we mean the process of inferring the metabolism of an organism from its genetic sequence data supplemented by known biochemical and phenotypic data. Our initial software system to support metabolic reconstruction was called WIT (for "What Is There?") and has been in use since mid-1995 (http://www.cme.msu.edu/ WIT/) [7]. Recently, a second system, which we have called WIT2, has been made available (http://www.mcs.anl.gov/home/overbeek/WIT2/CGI/ user.cgi). In this chapter we discuss the central design issues in constructing such systems, along with the basic steps that must be supported by any such system.

Representation of Metabolism

The most basic decisions center on how to represent the metabolism of an organism. Clearly, a topic of such complexity might well warrant an extremely abstruse computational representation. Indeed, the efforts that have been spent in representing chemical compounds give some indication of the potential magnitude of the problem.

In considering this problem, we have found it useful to draw an analogy to the representation of an automobile as it appears in any auto parts store. In this context, the auto overview and parts catalog give an accurate, high-level abstraction that does not include any real discussion of the "intermediates". It is an effective representation, but it does not convey the details of how energy is generated and distributed, how control mechanisms function, or how dynamic behavior is constrained.

We have developed an approach for representing the metabolism of an organism that is based on similar simplifications:

- We begin with a set of metabolic pathway diagrams. For our purposes, these diagrams are an arbitrarily structured and complex representation of a functional subsystem. Hence, we call them *function diagrams*. Just as an abstract drawing in an auto parts catalog attempts to convey the essential relationship of a set of functionally related parts, the function diagrams that we use attempt to convey the functional grouping of a set of proteins (normally, the set of enzymes that catalyze the reactions depicted in a metabolic pathway).

- Function diagrams themselves can (and should be) a well-structured representation of the functional interactions of proteins. This will be critical to support systems that base computations on the details of the interactions. For the purpose of metabolic reconstruction, however, none of this is necessary. A minimal function diagram composed simply of a list of protein identifiers would work just as well (i.e., we could use a set of minimal function diagrams in which each diagram was nothing but a list of enzymes along with any additional noncatalytic proteins).

- The central issue in this highly simplified framework now becomes how to assign unique identifiers to the *functional roles* in the diagrams. For function diagrams describing metabolic pathways, the enzyme number is usually adequate. However, some enzyme numbers are imprecise (i.e., they describe a class of enzymes), and there is the issue of what identifier to use for noncatalytic functional roles. As a (not completely adequate) solution, we use a slightly distilled version of the Swiss Protein Data Bank descriptions. (The people maintaining the Swiss Protein Data Bank have been making heroic efforts to standardize descriptions of protein functional roles, and whenever possible we simply exploit their efforts.)

The initial set of function diagrams that we now include in WIT/WIT2 comes from the *Metabolic PathwayDdatabase* built by Evgeni Selkov [8]. It now contains well over 2500 pathways and variants of pathways. These have been supplemented by a much smaller set of additional function diagrams from other sources.

Another way to summarize our representation of functional groupings is to say that we begin with two relational tables: (1) the *diagram-role table*, which contains two columns: the diagram identifier and the functional role identifier; and (2) the *protein-role table*, which also contains two columns: the protein sequence identifier and the functional role identifier .

Swiss Protein Data Bank entries are one class of protein sequences, and for them the "protein sequence identifier" is just the accession number. When other classes of protein sequence are used (e.g., ORFs from a newly sequenced genome), appropriate identifiers are used.

A metabolic reconstruction for a genome amounts to the entries in the protein--role table corresponding to ORFs from the genome, and a third table, the *asserted-diagrams table*, which is a list of the diagrams that have been asserted for the genome.

We stress that our approach of using arbitrary function diagrams and treating them as no more than collections of functional roles is a critical simplification. Such a simplification makes it possible to proceed with our goal of creating metabolic reconstructions without facing the detailed issues required to make inferences about the metabolic network. At the same time, if the actual function diagrams are a well-structured representation of the functions, such inferences will become commonplace (and useful in supporting the derivation and analysis of metabolic reconstructions).

How Is Metabolic Reconstruction Done?

Once the ORFs for a newly sequenced genome have been determined [1,2], we must carry out four steps: (1) assign functional roles, (2) assert the functional diagrams, (3) determine missing functions, and (4) balance the model.

Initial Assignments of Function

Our first step is to make initial assignments of their functional roles. This is done in two substeps: first, assignments are automatically generated for cases in which there appears to be relatively little ambiguity, and second, a manual pass through the ORFs with strong similarities but no assigned function is made.

Techniques for automatically assigning functional roles are advancing rapidly. We currently use the following approach for a translated ORF x from genome $g1$:

1. Compute similarities between the ORF and all sequences in the nonredundant protein sequence database. Save those above some designated threshold.
2. Consider similarities against ORFs from a completely sequenced genome $g2$. If x is similar to y from $g2$, and y is the protein in $g2$ closest to x, we say that y is a *best hit* (BH) against x. If x is also the best hit in - $g1$ against y, then we say that y is a *bidirectional best hit* (BBH).
3. Collect the set of BBHs for x. If the functional roles already assigned to those BBHs are all identical, assign the same functional role to x.

This is a quite conservative approach, although it can still lead to errors. Following the automated assignment of function, we recommend that the user of WIT/WIT2 make a pass through the set of ORFs that have strong similarities to other proteins with known functional roles but for which no automated assignment could be made. WIT2 allows the user to peruse the BBHs for each protein, to align the protein against other proteins of known function, to analyze regions of similarity, and so forth. At this point, assignment of function is still a process of thoughtfully considering a wide range of alternatives, and the background of the user determines the quality of the assignments. We believe that the rapid addition of new genomes and the accumulation of a growing body of probable assignments of function, together with consistency checks based on clustering protein sequences, will lead to a situation in which most of the currently required judgment can be eliminated. However, we are not yet close to that point.

An Initial Set of Pathways

Once the initial assignment of functional roles has been completed (i.e., once the initial version of the entries in the protein-role table for the newly sequenced genome has been generated), one normally proceeds to the assertion of function diagrams (i.e., to the addition of entries to the asserted-diagrams table for the genome). As the collection of analyzed genomes increases, it becomes ever more likely that each new genome will contain a substantial similarity to a genome that has already been analyzed. If a fairly similar (biochemically and phenotypically) organism has already been analyzed, it is useful to begin the analysis of the new organism by asserting the diagrams that are believed to exist from the already analyzed organism. Some of the asserted pathways are likely to be wrong, but their removal can be deferred until after the initial assignment of pathways.

In any event, the user should move through the major areas of metabolism and ask the system to propose diagrams that might correspond to functionality present in the organism. A system supporting metabolic reconstruction should be able to support such requests. As we learn more about the reasoning required to accurately assert the presence of pathways, the proposal of pathways by the system can become increasingly precise. For now, we employ a very straightforward approach.

First, we take the entire collection of pathways and assign a score to each pathway. The score for a pathway is

$$(I + 0.5U) / (I + U + M),$$

where I is the number of functional roles in the diagram that have been connected to specific sequences in the genome, M is the number that have not been connected and for which known examples from other genomes exist, and U is the number of unconnected roles for which no exemplar exists from other genomes. This is a crude measure of the fraction of the functional roles that have been identified, considering that there are U roles for which reasoning by homology is impossible at this point.

Then, we sort the pathways by score and present to the user those that exceed some specified threshold. The user is expected to go through each proposed pathway and either assert it to the asserted-diagrams table or simply ignore the proposal.

Locating Missing Functions

After we have accumulated an initial set of asserted diagrams, a pass through this asserted set must be made, focusing on the functional roles that remain unconnected to specific ORFs in the genome. Here, the system can provide a very useful function by collecting all known sequences that have been assigned the functional role, tabulating all similarities between ORFs in the new genome and these existing exemplars, and summarizing which of the existing ORFs is most likely to perform the designated functional role. Without a tool like WIT/WIT2, this process would be extremely time-consuming (and, in fact, would almost never be done systematically). In WIT2, we made the design decision to precompute similarities between all ORFs from the analyzed genomes and between these ORFs and entries in the nonredundant protein database maintained by NCBI. This allows an immediate response to requests to locate candidates for unconnected functional roles, summarizing BHs, BBHs, and all other similarities. The disadvantage of such a design commitment is that the collection of similarities is out of date almost immediately. Such a trade-off is commonly faced in developing bioinformatics servers. In our case, the severity of the problem is inevitably reduced by the addition of more genomes – that is, while the system may well not have access to all relevant similarities, the chances of establishing a solid connection between a new sequence and a previously analyzed sequence with an established function improve dramatically as the set of completely sequenced (and increasingly analyzed) genomes grow.

Once the system has located candidates for an unconnected functional role, the process of actually coming to a conclusion about whether a given sequence should be connected to the functional role is arbitrarily complex and corresponds to the types of decisions made while doing the initial assignments. In this case, however, the user of the system has the additional knowledge that assignments based on weak similarities may be strongly supported by the presence of assignments to other functional roles from the same diagram. This represents one of the pragmatic motivations for developing metabolic reconstructions: they offer a means of developing strong support for assignments based on relatively weak similarities.

We emphasize that the assertion of specific diagrams (i.e., pathways) should be considered in the context of known biochemical and phenotypic data. A variety of assignments cannot be made solely based on sequence similarities. For example, one might consider the choice between malate dehydrogenase and lactate dehydrogenase. Although examples of sequences that play these roles are extremely similar (exhibiting almost arbitrarily strong similarity scores), the choice between these functional roles often can be made only by using biochemical evidence or a more detailed sequence analysis based on either the construction of trees or the analysis of "signatures" (i.e., positions in the sequence that correlate with the functional role). Similarly, the choice between assigning a functional role of aspartate oxidase, fumarate reductase, or succinate dehydrogenase will require establishing an overview of the lifestyle of the organism, followed by a detailed analysis of all related sequences present in the genome. These examples are unusually difficult; in most cases the determination of function is much more straightforward. Even in these cases, however, the accumulation of more data will dramatically simplify things.

Balancing the Model

We turn now to the more difficult and critical step of balancing the model. By *balancing*, we mean considering questions of the following form:

"Since we know this compound is present (because we have asserted a given pathway for which it is a substrate), where does it come from? Is it synthesized, or is it imported?"

This consideration holds for all substrates to pathways, coenzymes, prosthetic groups, and so forth. In addition, we need to consider the issue of whether products of pathways are consumed by other cellular processes or are excreted.

To begin this process, the user must first make tables including all substrates of asserted pathways and all products of asserted pathways. As we stated above, our simplified notion of function diagram does not require that substrates and products be included. However, if one wishes to automate this aspect of metabolic reconstruction (which we have not yet done), the data must be accurately encoded. Once such tables exist, we can remove all compounds that occur as both substrates and products. Two lists remain:

1. A list of substrates that are not synthesized by any process depicted in any of the asserted function diagrams, and
2. A list of products that are not consumed by an processes depicted by asserted diagrams.

The user must go through these lists carefully and assess how best to reconcile the situation. This task may require searching for a protein that might be a potential transporter, asserting a new pathway for which a limited amount of evidence exists, or formulating some other hypothesis about what is going on.

Once the user has analyzed the situation as it relates to substrates and products of pathways, a similar analysis must be applied to known cofactors, coenzymes, and prosthetic groups. In this case, the logical issue of potential producers and consumers of specific compounds must be analyzed, but additional issues relating to volumes of flows can be analyzed. At this point, most of this type of analysis requires a substantial amount of expertise, and many of the decisions are necessarily impossible to make with any certainty. The situation is exacerbated by the difficulty of determining the precise function of a wide class of transport proteins, as well as by the potential for broad specificity for many enzymes. In this regard, while the situation is currently tractable only for those with substantial biochemical backgrounds (and not always by them), it is clearly possible that rapid advances in our ability to perform more careful comparative analysis and to acquire biochemical confirmation of conjectures will gradually simplify this aspect of metabolic reconstruction, as well.

Coordinating the Development of Metabolic Reconstructions

A metabolic reconstruction can be done by a number of individuals, often sharing a single model that is developed jointly. WIT2 includes the capability for multiple users either to work jointly on a single metabolic reconstruction or to develop such reconstructions in isolation. This is achieved as follows:

- For each organism, a list of *master users* is installed. When these users alter a model, the change is visible by all users of the system.
- When a user logs into a version of WIT2, he chooses a "user ID". Any set of users sharing the same ID will be working on the same model.
- When any non-master user alters a model (asserts the existence of a diagram or makes an entry to the protein-role table), the change is visible only to the group of users sharing the same user ID. The model constructed within a given user ID should be viewed as an extension to the "standard" model generated by the master users.
- A metabolic reconstruction for an organism (corresponding to a designated user ID) can be exported (i.e., converted to an external format), which can later be imported to any other version of WIT2 that includes the data for the organism.

Our intent is that users develop metabolic reconstructions on many distinct Web servers, but that they be able to conveniently import the efforts of others working on the same genome.

Where Do We Stand?

At this point we are attempting to develop and maintain metabolic models for well over twenty organisms representing a remarkable amount of phylogenetic diversity (http://wit.at.msu). The development of these initial models will be, we believe, far more difficult than the efforts required to add new models for more organisms that are similar to these initially analyzed organisms. On the other hand, unicellular life exhibits an enormous amount of diversity; and when the task of analyzing multicellular organisms is contemplated, it is clear that an enormous amount of work is required to attain even approximate metabolic reconstructions.

As we develop these initial models, we have noted a clear core of functionality that is shared by a surprisingly varied set of organisms. Techniques for developing clusters of proteins that are clearly homologous and that perform identical functions in distinct organisms are now beginning to simplify efforts to develop metabolic reconstructions. Such techniques are also leading to a clear hypothesis about the historical origins of specific functions.

The task of constructing a detailed overview of the functional subsystems in specific organisms is closely related to the issue of characterizing the functions or genes in the gene pool. While specific organisms often have been analyzed in isolation, it is rapidly becoming clear that comparative analysis is the key to understanding even specific genomes and that characterization of the complete gene pool for unicellular life is far more tractable than previously imagined. Our goal is to develop accurate, although somewhat imprecise, functional overviews for unicellular organisms and to use these as a foundation for the analysis of multicellular eukaryotes. Just as protein families derived from unicellular organisms are beginning to form the basis for assigning function to many eukaryotic proteins, an understanding of the central metabolism of eukaryotes will be built on our rapidly expanding understanding of the evolution of functional systems within unicellular organisms.

A Growing Interest in Connecting Metabolic and Sequence Data

The growing perception that the metabolic structure must be encoded and used to interpret the emerging body of sequence data has resulted in a number of projects. Here we summarize the most successful of these projects at this time. With interest expanding so rapidly, the reader is encouraged to do a network search for other sites, which we believe will continue to appear at a growing rate.

- KEGG (http://www.genome.ad.jp/kegg/kegg3.html) [4]: This outstanding effort, based at Kyoto University in Japan, represents an attempt to maintain metabolic overviews for sequenced genomes. It has connected the genes from specific organisms to metabolic functions with excellent visual depictions of metabolic maps.

- Boehringer Manheim Biochemical Pathways (http://expasy.hcuge.ch/cgi-bin/search-biochem-index): This excellent collection of metabolic pathways has been recently integrated into the SwissProt effort, allowing one to move between pathways, enzymes, and sequence data.

- EcoCyc (http://www.ai.sri.com/ecocyc/ecocyc.html - Overview) [5]: This database is a detailed encoding of the metabolism of Escherichia coli and Haemophilus influenzae. Besides just the metabolic network, this collection includes some of the kinetic and thermodynamic parameters (when they are known).

- Biocatalysis/Biodegradation Database (http://dragon.labmed.umn.edu/~lynda/index.html) [3]: This database covers a small, but significant, set of pathways that are of special interest in the area of xenobiotic degradation.

- SoyBase (http://probe.nal.usda.gov:8000/plant/aboutsoybase.html): This databases captures genetic and metabolic data for soybeans.

- Maize DB (http://teosinte.agron.missouri.edu/) [6]: This database is a comprehensive collection of maize genetic and biochemical data.

Availability of the Pathways, Software, and Models

The PUMA (http://www.mcs.anl.gov/home/compbio/PUMA/Production/puma. html), WIT (http://www.cme.msu.edu/WIT/) [7], and WIT2 (http://www.mcs.anl.gov/home/overbeek/WIT2/CGI/user.cgi) systems were developed at Argonne National Laboratory in close cooperation with the team of Evgeni Selkov in Russia. The beta release for WIT2 has been sent to four sites and is currently available. The first actual release of WIT2 is scheduled for October 1997. It will include all of the software required to install WIT2 and develop a local Web server, all of our metabolic reconstructions for organisms with genomes in the publicly available archives, and detailed instructions for adding any new genomes to the existing system (perhaps, for local use only). Just as widespread availability of the Metabolic Pathway Database has stimulated a number of projects relating to the analysis of metabolic networks, we hope that the availability of WIT2 will foster the development and open exchange of detailed metabolic reconstructions.

Acknowledgments

R.O. was supported by the U.S. Department of Energy, under Contract W-31-109-Eng-38. N.L. was supported by the Center for Microbial Ecology at Michigan State University (DEB 9120006). We also thank the Free Software Foundation and Larry Wall for their excellent software.

References

1. Badger, J. H., and Olsen, G. J. CRITICA: Coding Region Identification Tool Invoking Comparative Analysis, Molec. Bil. Evol., 1977, in press.
2. Borodovsky M., and Peresetsky, A. Deriving Non-Homogeneous DNA Markov Chain Models by Cluster Analysis Algorithm Minimizing Multiple Alignment Entropy. Comput Chemistry, 18, no. 3, 1994, pp. 259-267.
3. Ellis, L.B.M., and Wackett, L. P. A Microbial Biocatalysis Database, Soc. Ind. Microb. News. 45, no. 4, 1995, pp. 167-173.
4. Kanehisa, M., Toward Pathway Engineering: A New Database of Genetic and Molecular Pathways, Science and Technology Japan, 59, 1996, pp. 34-38.
5. Karp, P, Riley, M., Paley, S., and Pellegrini-Toole, A. EcoCyc: Electronic Encyclopedia of E. coli Genes and Metabolism, Nucleic Acids Research, 25, no. 1, 1997
6. Nelson, O., Coe, E., and Langdale, J., Genetic Nomenclature Guide. Maize. Trends Genet., March, 1995, 20-21
7. Overbeek O., Larsen, N., Smith, W., Maltsev, N., and Selkov, E.. Representation of Function: The Next Step. Gene-COMBIS (on-line): 31 January 1997; Gene 191, no. 1,: GC1-9
8. Selkov, E., Basmanova, S., Gaasterland, T., Goryanin, I., Gretchkin, Y., Maltsev, N., Nenashev, V., Overbeek, R., .Panushkina, E., Pronevitch, I., Selkov Jr., E.., and Yunus, I. The Methabolic Pathway Collection from EMP: The Enzymes and Metabolic Pathways Database. Nucleic Acids Research, 24, no. 1 (database issue), 1996, pp. 26-29.

4 ECOCYC: THE RESOURCE AND THE LESSONS LEARNED

Peter D. Karp* and Monica Riley**

*SRI International, 333 Ravenswood Avenue, Menlo Park
CA 94025, USA, pkarp@ai.sri.com
**Marine Biological Laboratory, Woods Hole, MA 02543,
mriley@mbl.edu

Introduction

The EcoCyc DB has several organizing principles. It is organized around the bacterium *E. coli* K—12. It is organized at the level of a review in that a given entry in the DB encodes information from a variety of sources about a single biological entity, such as an enzyme. A former organizing principle of the DB was to focus on information about enzymes and metabolic pathways; however, that focus is broadening to include transport, regulation, and other aspects of gene function.

EcoCyc is more than a DB – it is also a suite of software tools for visualizing and querying genomic and metabolic data. This chapter describes both the DB and the software tools. It surveys the content of the DB, and the mechanisms by which new data are acquired and validated. We close by discussing some of the lessons learned from the EcoCyc project.

The EcoCyc Data

The EcoCyc DB describes the known genes of *E. coli*, the enzymes of small-molecule metabolism that are encoded by these genes, the reactions catalyzed by each enzyme, and the organization of these reactions into metabolic pathways. The EcoCyc graphical user interface software (GUI) allows scientists to query, explore, and visualize the EcoCyc DB. EcoCyc therefore integrates both genomic data and detailed descriptions of the functions of gene products. The EcoCyc data were drawn largely from (and contain 1650 citations to) the primary literature. In addition, some data were obtained from other DBs.

EcoCyc has potential uses in addition to its role as a reference source on *E. Coli*. Because of its links to sequence DBs such as Swiss-Prot, EcoCyc could be used to perform function-based retrieval of DNA or protein sequences, such as to prepare datasets for studies of protein structure--function relationships. Scientists who study evolution of the metabolism could use EcoCyc to search out examples of duplication and divergence of enzymes and pathways. EcoCyc provides a quantitative foundation for performing simulations of the metabolism, although it currently lacks the quantitative kinetics data needed by most simulation techniques.

EcoCyc has been used to predict the metabolic complements of *H. pylori* [5] and of *H. influenzae* from their genomic sequences [15]. The latter metabolic prediction was materialized in DB form and combined with the EcoCyc software to create an encyclopedia of the *H. influenzae* genome, called HinCyc. This metabolic-analysis technique extracts an added level of biological information from a genomic sequence, and provides a biological validation of the gene identifications predicted by sequence analysis.

Biotechnologists seek to design novel biochemical pathways that produce useful chemical products (such as pharmaceuticals), or that catabolize unwanted chemicals such as toxins. EcoCyc provides the wiring diagram of *E. coli* K—12, which approximates the starting point for engineering; EcoCyc also describes the potential engineering variations that can result from importing *E. coli* enzymes into other organisms.

The EcoCyc Graphical User Interface

The EcoCyc GUI [7] provides graphical tools for visualizing and navigating through an integrated collection of metabolic and genomic information (its retrieval capabilities are described in [13]). For each type of biological object in the EcoCyc DB, the GUI provides a corresponding visualization tool. There are tools for visualizing pathways, reactions, compounds, and so forth. These tools dynamically query the underlying DB for one or more objects and produce drawings specific to those objects. All display algorithms are parameterized to allow the user to select the visual presentation of an object that is most informative. For example, the algorithms that produce automatic layouts of metabolic pathways can suppress the display of enzyme names or side-compound names; they can also draw chemical structures for the compounds within a pathway [9].

Organization of the EcoCyc Data

The EcoCyc data are stored within a frame knowledge representation system (FRS) called Ocelot (its capabilities are similar to those of HyperTHEO, described in [10]). FRSs use an object-oriented data model, and have several advantages over relational DB management systems [6]. FRSs organize information within *classes*: collections of objects that share similar properties and attributes. The EcoCyc schema is based on the class hierarchy shown in Figure 1 [12]. All the biological entities described in EcoCyc are instances of the classes in Figure 1. For example, each *E. coli* gene is represented as an instance of the class Genes, and every known polypeptide is an instance of the class Polypeptides.

We believe that the current version of EcoCyc contains all known enzymes and pathways of *E. coli* small-molecule metabolism (we expect that more enzymes will be discovered as the sequence of the *E. coli* genome is further analyzed). Table 1 lists the number of instances within the EcoCyc DB of selected classes in Figure 1

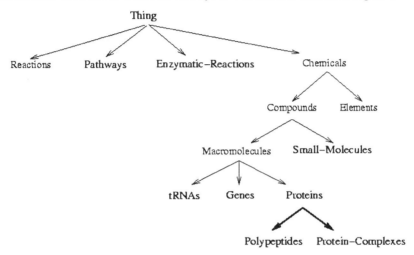

Figure 1: The top of the class hierarchy for the EcoCyc DB. The arrows in this figure point from a general class of objects to a more specific class of objects; for example, we divide the class Proteins into the subclasses Polypeptides and Protein-Complexes.

Class	Size
Reactions	595
Enzymes	695
Pathways	123
Genes	3030
tRNAs	79
Compounds	1296

Table 1: The number of objects in several EcoCyc classes. The Enzymes row gives the number of polypeptides or protein complexes that catalyze a reaction. The numbers for Polypeptides and Protein Complexes also include some transport proteins.

Each EcoCyc frame contains *slots* that describe attributes of the biological object that the frame represents, or that encode a relationship among that object and other objects. For example, the slots of a polypeptide frame encode the molecular weight of the polypeptide, the gene that encodes it, and its cellular location.

The scope of the data within EcoCyc is slowly expanding over time. The DB now describes most known *E. coli* genes; it describes those gene products that are enzymes involved in small-molecule metabolism, or that are tRNA synthetases, or that are involved in two-component signal transduction systems. It also describes gene products that are tRNAs. In the near future, we are planning to add descriptions of gene products that are transport proteins, or regulatory proteins.

New information is entered into EcoCyc using a combination of graphical editing tools. Some of these tools are specialized for entry of metabolic data, including graphical editors for reactions, compounds, and pathways. In addition, a domain-independent KB browsing and editing tool called the GKB Editor allows interactive editing of the EcoCyc class hierarchy, of a semantic-network view of the EcoCyc KB, and of individual EcoCyc frames; it also allows EcoCyc data to be transferred to a spreadsheet [17, 19].

Data validation techniques used in EcoCyc are described in [13].

We next describe the major classes of information within EcoCyc, the sources from which the information was obtained, and the visualization tools associated with those classes.

Genes

Most information on *E. coli* genes in EcoCyc was obtained from the EcoGene DB version 7 [2]. In the near future that information will be superseded by information from the full *E. coli* DNA sequence [3]. EcoGene provides synonyms for gene names, physical map positions for all sequenced genes, and the direction of transcription for each gene. We supplemented the information in EcoGene significantly by adding descriptions of additional *E. coli* genes obtained from the literature and from SwissProt. EcoCyc contains 3030 genes, of which 2571 have assigned genomic map positions. The map positions in EcoCyc version 3.7 were obtained from the EcoGene DB, but in the near future we wil obtain map positions from the full *E. coli* genomic sequence [3]

The visualization tool that generates gene-display windows lists information such as the map position of the gene on the *E. coli* chromosome (in units of centisomes, or hundredths of a chromosome), the class(es) to which the gene was assigned, and the direction of transcription. The gene product is listed (when known); when the product is an enzyme known to EcoCyc, the display shows the equation(s) of the reaction(s) catalyzed by the enzyme, and the pathways that contain those reactions.

We have classified EcoCyc genes according to two different classification systems. The first is based on the physiological role of the gene product (e.g., all genes whose products are involved in tryptophan biosynthesis, including enzymes and regulatory proteins, are in a single category) [20]. The second system is coarser, and assigns each gene product to one of the following classes: Enzymes, Regulators, Leaders, Membrane Proteins, Transporters, Structural Proteins, RNAs, Factors, Carriers, and products of unknown function.

The Gene--Reaction Schematic

The many-to-many relationships among genes, enzymes, and reactions can be complex. An enzyme composed of several subunits might catalyze more than one reaction, and a given reaction might be catalyzed by multiple enzymes. The *Gene— Reaction Schematic* depicts the relationships among a set of genes, enzymes, and reactions (see Figure 2). It is generated by starting with the object that is the focus of the current window (which is highlighted in the schematic), and then recursively traversing KB relationships from that object to related objects, such as from a gene to its product, or from a reaction to the enzyme(s) that catalyzes it. The schematic summarizes these complex relationships succinctly, and also constitutes a navigational aid—the user can click on an object in the schematic to cause EcoCyc to display that object.

The first schematic in Figure 2 means that the trpA gene encodes a polypeptide (the circle to the left of the box for the trpA gene) that forms a heterotetramer (the next circle to the left —the 2 indicates two copies) that also contains two copies of the product of the trpB gene. That complex in turn catalyzes reaction 4.2.1.20. The second schematic (reading down the column) depicts three isozymes (two homodimers and a homotetramer) that catalyze reaction 4.2.1.2. The third schematic depicts a bifunctional polypeptide, and the fourth schematic depicts a case where a homodimer of the TrpD polypeptide catalyzes one reaction, and a heterotetramer of TrpD and TrpE catalyzes a second reaction. The fourth schematic depicts two isozymes that each are heterotetramers. The fifth schematic depicts the ATP synthase protein, which consists of a large complex containing two subcomplexes.

Schematics also include modified forms of a protein (or tRNA) when relevant. For example, the schematic for the acyl carrier protein shows both a yellow circle for the unmodified form of the protein, and 13 orange circles, which represent different modified forms of the protein.

52

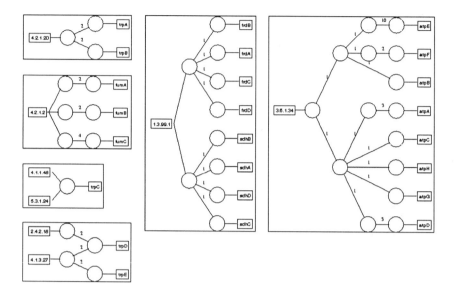

Figure 2: A set of gene--reaction schematics. Some of the reactions in the largest schematic have no assigned EC number. The boxes to the left represent reactions, the boxes on the right represent genes, and the circles in the middle represent proteins. The lines indicate relationships among these objects. The schematic is drawn in gene windows, reaction windows, and protein windows.

Reactions

The reactions within EcoCyc were gathered from biomedical literature on *E.coli.* In addition, we incorporated many non-*E. coli* reactions and 269 reaction classes that constitute the enzyme classification system [21] from the ENZYME DB [1]. EcoCyc therefore contains many reactions not found in *E.coli*, for reference purposes. EcoCyc reaction windows state whether or not we have evidence that a given reaction occurs in *E.coli*.

The reaction display window shows the class(es) containing the reaction within the classification of reactions. It shows the one or more enzymes that catalyze the reaction, the gene(s) that code for the enzymes, and the pathway that contains the reaction. The display shows the EC number for the reaction, and the reaction equation. Note that there exists a one-to-one mapping between EC numbers and reactions, but not between EC numbers and enzymes [11], therefore, we label reactions, and not enzymes, with the EC number. The standard change in Gibbs free energy of the reaction is listed when known.

Proteins

EcoCyc contains extensive information about *E. coli* enzymes and pathways that we obtained from the biomedical literature. We performed a comprehensive literature search for each *E. coli* enzyme, reaction, and pathway by using Medline, the *E. coli--Salmonella* book [16], and biochemistry textbooks. We searched for other pertinent papers by following citations in journal articles and in the *Science Citation Index.*

In the EcoCyc schema, all enzyme objects are instances of the class of all proteins, which we call `Proteins`; it is partitioned into two subclasses: `Protein-Complexes` and `Polypeptides`. These two classes have several common properties, such as molecular weight (when the stoichiometry of the protein-complex is known), cellular location, and a link to any reactions catalyzed by the protein. They differ in that `Protein-Complexes` have slots that link them to their subunits, whereas `Polypeptides` have a slot that identifies their gene. We record whether sequence-similarity relationships exist among a set of isozymes, and we provide links to the SwissProt, PDB, and Swiss-Model entries for a polypeptide. Proteins are listed as a subclass of chemicals since in some cases proteins themselves are substrates in a reaction (such as phosphorylation reactions).

For each enzyme, we have written comments that address topics such as reaction mechanism, subreactions of complex reactions, interactions of subunits of complex enzymes, formation of complexes with other proteins, breadth of substrate specificity, mode of action of inhibitors and activators, place and function of reactions in metabolic pathways, other reactions catalyzed by the protein, and relationship of the protein to other proteins catalyzing the same reaction.

Protein windows[1] are complicated because of the many-to-many relationship between enzymes and reactions (one enzyme can catalyze multiple reactions, and each catalytic activity of an enzyme can be influenced by different cofactors, activators, and inhibitors), and because many genes can encode the subunits of a protein complex. The protein window is potentially divided into sections to address these complexities.

The first section of the window lists general properties of the protein, such as synonyms and molecular weight. Subsequent sections of the window describe each catalytic activity of the protein, if it is an enzyme. Each activity section lists a reaction catalyzed by the enzyme, and the enzyme name (and synonyms) for that activity. The substrate specificity of the enzyme is described in some cases by listing alternative compounds that the enzyme will accept for a specified substrate. The cofactor(s) and prosthetic groups required by the enzyme are listed next,[2] along with any known alternative compounds for a specified cofactor. Activators and inhibitors of the enzyme are listed, qualified as to the mechanism of action, when known. In addition, this section indicates which of the listed activators and inhibitors are known to be of physiological relevance, as opposed to whether the effects are known purely because of *in vitro* studies. For a multifunctional nzyme, the descriptions of substrate

---

[1] See URL http://ecocyc.ai.sri.com:1555//NEW IMAGE?type=ENZYME\&object= LACTALDREDUCT-CPLX for an example.

specificity, cofactors, activators, and inhibitors are all tied to the enzyme activity to which they pertain. For more details on how this information is represented in EcoCyc, see [11].

Pathways

Pathway frames list the reactions that make up a pathway, and describe the ordering of those reactions within the pathway. Information about the ordering of reactions within a pathway is ncoded using a predecessor-list representation [8], which for each reaction in a pathway lists the reactions that precede it in the pathway. This representation allows us to capture complex pathway topologies, yet does not require entering information that is redundant with respect to existing reaction objects. We developed algorithms for deriving a graph description of the pathway from the predecessor list [8].

The DB uses objects called superpathways to define a new pathway as an interconnected cluster of smaller pathways. For example, a superpathway called ``complete aromatic amino-acid biosynthesis'' links together the individual pathways for biosynthesis of chorismate, tryptophan, tyrosine, and phenylalanine. Superpathways are also defined using the predecessor list [8]. EcoCyc currently contains 123 pathways and 34 superpathways.

All pathway drawings in EcoCyc are computed automatically using pathway-layout algorithms (see Figure 3). EcoCyc can draw pathways at multiple levels of detail, ranging from a skeletal view of a pathway that depicts the compounds only at the periphery of the pathway and at internal branch points, to a detailed view that shows full structures for every compound, and EC numbers, enzyme names, and gene names at every reaction step. Users can select among these views by clicking on buttons labeled More Detail and Less Detail.

Compounds

The class Chemicals subsumes all chemical compounds in the *E. coli* cell, such as macromolecules and smaller compounds that act as enzyme substrates, activators, and inhibitors. It also includes some of the elements of the periodic table. Small metabolites contained in EcoCyc are reaction substrates, and enzyme cofactors, activators, and inhibitors.

EcoCyc contains 1294 compounds; two-dimensional structures are recorded for 965 of them. Among the properties encoded for compounds are synonyms for their names, molecular weight, empirical formula, lists of bonds and atoms that encode chemical structures, and two-dimensional display coordinates for each atom that permit drawings of compound structures.

A compound display lists all EcoCyc reactions in which the compound appears, sorted by the pathways that contain each reaction. The display of chemical structures within compound windows uses a concept called superatoms, which is a hierarchical structuring of chemical structures. For example, the structure for succinyl-CoA is initially displayed with the word ``CoA'' in place of the structure of the CoA moiety. If the user clicks on the word CoA, however, the full structure of that moiety is displayed.

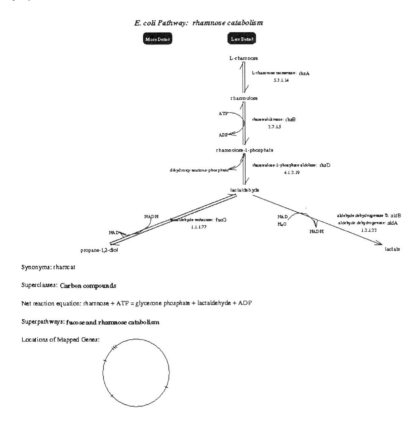

Figure 3: An EcoCyc drawing of the pathway for rhamnose catabolism. The small circle at the bottom of the pathway window depicts the positions on the *E. coli* genomic map of the genes that encode the enzymes within the current pathway. When the user moves the mouse over a given gene, its name is printed at the bottom of the EcoCyc window; clicking on the gene causes EcoCyc to display a window for that gene.

The Metabolic-Map Overview

The Overview diagram is a drawing of all known metabolic pathways of *E.coli*. In this diagram, each circle represents a single metabolite, and each line represents a single bioreaction. Neither the circles nor the lines are unique in the sense that a given metabolite (or a given reaction) may occur in more than one position in the diagram. The ``barbells'' along the right side of the diagram represent approximately 200 individual reactions that have not been assigned to a particular pathway. They are presented as single reactions because their direction and role are determined by the metabolic condition of the cell. (The barbell region also contains some reactions of macromolecule metabolism, such as tRNA charging and DNA metabolism.) In the pathway region to the left of the barbells, the glycolysis and the TCA cycle pathways in the middle separate predominately catabolic pathways on the right from pathways of anabolism and intermediary metabolism on the left. The existence of anaplerotic pathways prevents rigid classification. The pathway region contains 500 reactions organized into 123 pathways. The majority of those pathways operate in the downward direction.

A user can interrogate the Overview in several ways. To identify a compound within the Overview, the user moves the mouse pointer over a circle in the diagram — EcoCyc prints the name of the metabolite and the name of the containing pathway at the bottom of the screen. To identify a reaction, the user moves the mouse pointer over a line — the program prints the equation of that reaction, and the name of the containing pathway. If the user left-clicks on a compound or a reaction in the Overview, that object is displayed in its own display window. The user can also highlight objects in the Overview, such as finding a compound by name or finding a reaction by EC number.

Genomic-Map Displays

We have developed several methods for a scientist to visualize relationships among the roughly 1500 mapped *E. coli* genes in the EcoCyc KB. These methods are based on map display tools that allow the user to view the distributions of genes on a map at multiple resolutions.

The first visualization tool provides a linear depiction of the chromosome (see figure 4). Initially, a low-resolution view of the entire chromosome is available. The user can zoom in on a region of the chromosome in several ways. Middle clicking on any of the gene names shown in a higher-resolution drawing of the region of the chromosome centered on that gene. These high-resolution child maps can be generated to any desired level; that is, one can also click on a gene in a child map to produce yet more resolution. The user can also zoom by left clicking on the vertical line representing the chromosome, or by specifying a centisome position or a gene name. EcoCyc produces as many levels of child maps as are necessary to show the requested information.

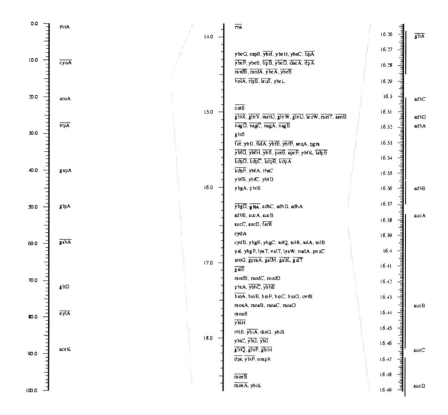

Figure 4: The EcoCyc linear map browser. Sections of the chromosome are shown at three resolutions. In the rightmost section, the coding region for each gene is shown with a vertical line.

The linear partial map allows an investigator to examine map relationships among subsets of *E. coli* genes by beginning with a display of the chromosome with no genes present, and selectively adding genes to the window. The genes to be added can be selected by several criteria: by name (such as *hisE*); by matching a substring against all gene names (such as ``his''); as members of one of the classes of all *E. coli* genes defined by Riley [20]; or the set of genes coding for enzymes in a chosen biochemical pathway.

The user can consecutively add sets of genes according to these criteria, and can undo previous add operations in reverse order. For example, the user might add all genes in the biosynthetic pathways for tryptophan, tyrosine, and phenylalanine, and then add all genes coding for membrane proteins. The zoom capabilities described in conjunction with the full map also work for the partial map.

The second tool displays the full map in circular form, with similar zooming capabilities.

Retrieval Operations

EcoCyc provides the user with two classes of DB retrieval operations: direct retrieval through menus of predefined queries, and indirect retrieval through hypertext navigation [13]. For example, imagine that a user seeks information on the *hisA* gene, such as its map position and information about the enzyme it encodes. EcoCyc allows the user to call up an information window for that gene directly by querying the gene name.

The indirect approach consists of hypertext navigation among the information windows for related objects. Such navigation allows the user to find the *hisA* gene by traversing many paths through the DB. The user could issue a direct query to display the biosynthetic pathway for histidine, and then click on the name of the enzyme at the last step in the pathway. The resulting information window for that enzyme shows the name of the gene (*hisA*) coding for the enzyme. Clicking on the gene name displays the information window for (*hisA*). Alternatively, the user could query the compound histidine by name. The resulting window lists all reactions involving histidine; the user can click on a reaction to navigate to its window, which lists all enzymes that catalyze the reaction, plus all genes encoding those enzymes (including *hisA*).

Software Architecture and Distribution

EcoCyc is implemented in Common Lisp with a graphical-interface toolkit called the Common Lisp Interface Manager (CLIM). The architecture of the EcoCyc development environment is shown in Figure 5. EcoCyc is available under license through the WWW, as a program that runs on the Sun workstation, and as a set of flat files. See the EcoCyc WWW pages at URL http://www.ai.sri.com/ecocyc/ for more information.

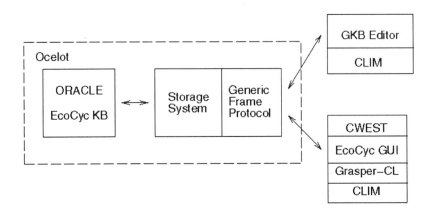

Figure 5: The software architecture of EcoCyc. The components include a graph-management system called Grasper-CL [14], an API for FRSs called the Generic Frame Protocol, and the CWEST tool for retrofitting CLIM applications to run through the WWW [18].

Lessons Learned from the EcoCyc Project

This section presents the bioinformatics contributions of the EcoCyc project and the lessons learned in the course of the project, and notes properties of EcoCyc that distinguish it from some other biological DBs.

One lesson learned is that metabolic-pathway DBs are a useful addition to the repertoire of biological DBs. They provide a reference source on metabolic pathways. They have been successfully used to predict the metabolic complements of organisms from their genomic sequence [15]. Other potential applications of these DBs include pathway design for biotechnology [14], and simulation of metabolic pathways.

It is useful to organize pathway databases as a collection of multiple types of biological objects: pathways, reactions, enzymes, genes, and compounds. This organization is to be contrasted with that of biological databases such as GenBank, Swiss—Prot, and PDB, which contain only a single type of biological object.

Metabolic information is complex. The EcoCyc project has explored issues in the representation of metabolic pathways, reactions, and enzymes, and has developed an ontology for metabolic information that could be reused by other metabolic DBs [11, 8]. Most past biological DBs have encoded biological function within English text fields of the DB, whereas we have developed structured, declarative representations of function with which the user can query and compute. We have emphasized the development and publishing of our ontology because of the complexities of representing enzyme function and metabolic pathways. However, we believe that other biological-DB projects would benefit from a similar emphasis. Publication of ontologies both increases the understanding of the DB in the user community and aids the developers of other similar DBs.

We have designed graphical presentations of metabolic information for the EcoCyc GUI, and algorithms for generating those presentations, including automated layout algorithms for metabolic pathways [9].

The knowledge-acquisition problem has been a serious concern throughout the EcoCyc project. The problem is that of translating information in the scientific literature into a set of interconnected frames within a KB. The problem is particularly severe because EcoCyc contains so many different object classes. The process of describing a single complex metabolic pathway could involve creating several dozen different instances of four different classes, all of which must be properly linked. We have addressed this problem in three different ways: we have developed pathway-specific graphical tools for entry of pathway, reaction, and enzyme information; we also employ a general KB browsing and editing tool that is not independent of biology; finally, we have extensively trained the project members who perform knowledge entry. The advantage of domain-independent knowledge acquisition tools is that they can be reused across application domains; however, for most tasks they cannot compete with the speed and intuitiveness of domain-specific tools.

Another advantage of graphical knowledge acquisition tools is that they can perform immediate checking of new information. In the first few years of our project, knowledge was entered into structured text files by one group of people, and then loaded into the KB by a second group of people who ran parsers over the files. The inevitable errors and inconsistencies in the information were time consuming for the two groups to resolve.

In its early phases, the EcoCyc project collected kinetics data for each metabolic enzyme. We ceased to collect this data upon discovering that the data are extremely time consuming to collect because there are large amounts of kinetics data in the literature, and upon discovering that this information is of relatively low utility because the data are collected under widely varying experimental conditions, and therefore cannot easily be combined to produce a coherent simulation of a set of enzymes under a single condition.

EcoCyc employs a frame knowledge representation system as its data management substrate. Its schema-evolution capabilities have simplified the development of the EcoCyc schema. Its object-oriented data model has yielded a much simpler and more comprehensible schema than would be possible under the relational model.

Development of the EcoCyc GUI began before invention of the WWW; therefore, EcoCyc was developed using a Common Lisp X-windows environment. We were able to retrofit EcoCyc to run through the WWW in addition to X-windows through the use of a tool called CWEST. CWEST translates low-level data structures produced by the Common Lisp windowing system into a combination of GIF images and HTML pages, and allowed us to adapt EcoCyc to the WWW with relatively little effort. Our use of a powerful windowing system was an extremely positive decision.

References

1. Bairoch. The ENZYME databank in 1995. *Nucl Acids Res*, 24:221--222, 1996.
2. Berlyn, K. Brooks Low, K.E. Rudd, and M. Singer. Linkage map of *Escherichia coli* K—12, edition 9. In Neidhardt et al. [16], pages 1715--1902.
3. Blattner, G. Plunkett III, C.A. Bloch, N.T. Perna, V. Burland, M. Riley, J. Collado-Vides, J.D. Glasner, C.K. Rode, G.F. Mayhew, J. Gregor, N.W. Davis, H.A. Kirkpatrick, M.A. Goeden, D.J. Rose, B. Mau, and Y. Shao. The complete genome sequence of *Escherichia coli* K—12. *Science*, 277:1453--1462, 1997.
4. Cameron and I. Tong. Cellular and metabolic engineering: An overview. *Applied Biochemistry and Biotechnology*, 38:105--140, 1993.
5. J.-F. Tomb et al. The complete genome sequence of the gastric pathogen *Helicobacter pylori*. *Nature*, 388:539--547, 1997.

6.   Karp. Frame representation and relational data bases: Alternative information-management technologies for systematics. In R. Fortuner, editor, *Advanced Computer Methods for Systematic Biology: Artificial Intelligence, Database Systems, Computer Vision*, page 560. The Johns Hopkins University Press, 1993.

7.   Karp. The EcoCyc user's guide.unpublished; see WWW URL `ftp://ftp.ai.sri.com/pub/papers/karp-ecocyc-guide.ps.Z`, 1996.

8.   Karp and S. Paley. Representations of metabolic knowledge: Pathways. In R. Altman, D. Brutlag, P. Karp, R. Lathrop, and D. Searls, editors, *Proceedings of the Second International Conference on Intelligent Systems for Molecular Biology*, pages 203--211, Menlo Park, CA, 1994. AAAI Press.

9.   Karp and S. Paley. Automated drawing of metabolic pathways. In H. Lim, C. Cantor, and R. Robbins, editors, *Proceedings of the Third International Conference on Bioinformatics and Genome Research}*, pages 225--238. World Scientific Publishing Co., 1995. See also WWW URL `ftp://ftp.ai.sri.com/pub/papers/karp-bigr94.ps.Z`.

10.   Karp and S. Paley. Integrated access to metabolic and genomic data. *Journal of Computational Biology*, 3(1):191--212, 1996.

11.   Karp and M. Riley. Representations of metabolic knowledge. In L. Hunter, D. Searls, and J. Shavlik, editors, *Proceedings of the First International Conference on Intelligent Systems for Molecular Biology*, pages 207--215, Menlo Park, CA, 1993. AAAI Press.

12.   Karp and M. Riley. Guide to the EcoCyc schema. unpublished; see WWW URL `ftp://ftp.ai.sri.com/pub/papers/karp-ecocyc-schema.ps`, 1996.

13.   Karp, M. Riley, S. Paley, A. Pellegrini-Toole, and M. Krummenacker. EcoCyc: Electronic encyclopedia of *E. coli* genes and metabolism. *Nuc. Acids Res.*, 25(1):43--50, 1997.

14.   Karp, J.D. Lowrance, T.M. Strat, and D.E. Wilkins. The Grasper-CL graph management system. *LISP and Symbolic Computation*, 7:245--282, 1994. See also SRI Artificial Intelligence Center Technical Report 521.

15.   Karp, C. Ouzounis, and S.M. Paley. HinCyc: A knowledge base of the complete genome and metabolic pathways of *H. influenzae*. In D.J. States, P. Agarwal, T. Gaasterland, L. Hunter, and R. Smith, editors, *Proceedings of the Fourth International Conference on Intelligent Systems for Molecular Biology*, pages 116--124, Menlo Park, CA, 1996. AAAI Press.

16.   Neidhardt, III R. Curtiss, J. Ingraham, E.C.C. Lin, K.B. Low, B. Magasanik, W. Reznikoff, M. Riley, M. Schaechter, and H. E. Umbarger, editors. *Escherichia coli and Salmonella, 2nd edition*. ASM Press, 1996.

17.   Paley and P. Karp. GKB Editor user manual. Available via WWW URL `http://www.ai.sri.com/~gkb/user-man.html`, 1996.

18.   Paley and P.D. Karp. Adapting EcoCyc for use on the World Wide Web. *Gene*, 172(1):GC43--50, 1996.

19. Suzanne M. Paley, John D. Lowrance, and Peter D. Karp. A generic knowledge-base browser and editor. In *Proceedings of the 1997 National Conference on Artificial Intelligence*, 1997.
20. Riley. Functions of the gene products of *Escherichia coli*. *Microbiological Reviews*, 57:862--952, 1993.
21. Edwin C. Webb. *Enzyme Nomenclature, 1992: Recommendations of the nomenclature committee of the International Union of Biochemistry and Molecular Biology on the nomenclature and classification of enzymes.* Academic Press, 1992.

5 KEGG: FROM GENES TO BIOCHEMICAL PATHWAYS

Minoru Kanehisa

Institute for Chemical Research, Kyoto University, Kyoto, Japan

Introduction

Molecular biology has been a discipline dominated by the reductionistic approach where starting from a specific functional aspect of a biological organism the genes and proteins that are responsible for the function are searched and characterized. In contrast, the complete set of genes and gene products that has become available by the whole genome sequencing is a starting point of an alternative approach, which may be called a synthetic approach, toward understanding how genes and molecules are networked to form a biological system. While it is unlikely that the reductionistic approach alone can cover the entire aspects of the biological system, the synthetic approach has a potential to provide a complete picture because the starting set of building blocks is complete. In reality, however, the complete genome sequence does not tell much about how the organism functions as a biological system. This is not only because we do not yet have appropriate means to interpret the sequence data, but also because all the information to build up a biological system may not be present in the genome.

KEGG (Kyoto Encyclopedia of Genes and Genomes) is an effort to make links from the gene catalogs generated by the genome sequencing projects to the biochemical pathways that may be considered wiring-diagrams of genes and molecules [1]. Specifically, the objectives of KEGG are the following:

1. to computerize all aspects of cellular functions in terms of the pathway of interacting molecules or genes,

2. to maintain gene catalogs for all organisms and link each gene product to a pathway component,

3. to organize a database of all chemical compounds in the cell and link each compound to a pathway component, and

4. to develop computational technologies for pathway comparison, reconstruction, and analysis.

KEGG is publicly made available as part of the Japanese GenomeNet Service [2]. The WWW addresses that are relevant to this chapter are summarized in Table 1.

Table 1. The Japanese GenomeNet service

Service	URL
GenomeNet	www.genome.ad.jp
DBGET/LinkDB	www.genome.ad.jp/dbget/
	www.genome.ad.jp/dbget/dbget.links.html
KEGG	www.genome.ad.jp/kegg/
	www.genome.ad.jp/kegg/kegg2.html
LIGAND	www.genome.ad.jp/dbget/ligand.html
BRITE	www.genome.ad.jp/brite/

Data Representation in KEGG

Level of abstraction

As illustrated in Figure 1, a cysteine is a network of carbon, nitrogen, oxygen, hydrogen, and sulfur atoms at the atomic level, but it is abstracted to letter C at the molecular level where a protein is represented by a one-dimensional network of twenty letters (amino acids). In the next network level the protein is abstracted to a symbol, Ras, and the wiring among different symbols (proteins) is the major concern as in this case of Ras signal transduction pathway. Most of the current molecular biology databases contain the data at either the molecular level or the atomic level focusing on the information that is associated with protein and nucleic acid molecules, such as the sequence databases and the 3D structural databases. Consequently, the majority of the current computational methods in molecular biology involves comparison and analysis of sequence data or 3D atomic coordinate data. The main role of KEGG is to fill the gap at the network level by computerizing current knowledge of biochemical pathways and by developing associated computational technologies. KEGG also intends to fill another gap at the atomic level, which is the lack of good public resources for chemical compounds.

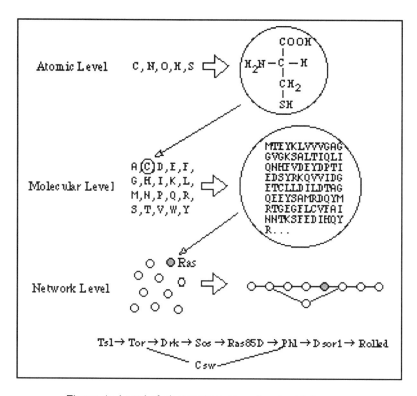

Figure 1: Level of abstractions in molecular biology.

Binary relations

An important concept in KEGG is the binary relation that represents any relation between two elements [3]. Of course, the real-world data can better be represented by considering ternary, quaternary, and higher relations, but for the sake of logical simplicity and computational efficiency KEGG imposes the view of binary relations. A major class of binary relation is an interaction between two molecules or between two genes, which may be considered wiring information of the biological system. Generally speaking, building blocks can contain all necessary wiring information as in the jigsaw puzzle, but in reality building blocks of life are more like Lego blocks that can be used in many different ways. KEGG computerizes the information of interactions and wiring, thus complementing the existing databases for the information of building blocks.

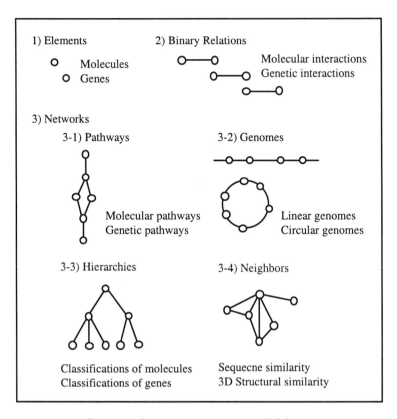

Figure 2: Data representation in KEGG.

Different types of networks

Figure 2 illustrates the data hierarchy in KEGG that consists of elements, binary relations, and networks, and also examples of different types of networks. A network is any type of relationship among a set of elements. For example, the biochemical pathway is a network of interacting molecules and the genome is a one-dimensional network of genes. The information of neighbors that can be computed by sequence similarity search or 3D structural similarity search among a set of molecules also forms a complex network. The similarity relationship is often better organized in the form of a hierarchy, such as the superfamily classification or the 3D fold classification, which is still another type of network.

Table 2 summarizes the actual implementation of different types of networks in KEGG. The biochemical pathway is represented by the graphical pathway map, which is handled as a GIF image map by HTML, CGI scripts, and Java applets. The genomic information is represented by the gene catalog that is a functional hierarchy of genes and the graphical genome map. The gene catalog as well as the catalog of proteins and other molecules is handled by the CGI script for, what we call, hierarchical text. The genome map is handled by Java applets. The sequence similarity relationships of genes within and across species are analyzed in KEGG to identify functional relationships. Thus, the information of orthologous and paralogous genes is represented in the ortholog group table, which is a simple HTML table.

Table 2. Different types of network data in KEGG.

Category	KEGG data	Representation
Pathway	Pathway map	GIF image map
Genome	Genome map	Java graphics
Hierarchy	Gene/molecule catalog	Hierarchical text
Neighbor	Ortholog group table	HTML table

Database Management Systems

DBGET/LinkDB system

The DBGET/LinkDB system [4] is an in-house developed database management system written in C++ that provides the basis for integrating different resources within KEGG and also with other existing databases. DBGET/LinkDB integrates a diverse range of databases at the level of entries. An entry in the web of molecular biology databases is uniquely identified by the combination of the database name and the entry name (or accession number)

database:entry

and the relation between two entries within and across databases is represented by the binary relation:

database1:entry1 $\rightarrow$ database2:entry2

This concept has been extended to include gene catalogs in KEGG. Any gene (or gene product) in the web of biological organisms is uniquely identified by the combination of the organism name and the gene name

organism:gene

and the relation between two genes (or gene products) within and across organisms is represented by the binary relation:

$$\text{organism1:gene1} \rightarrow \text{organism2:gene2}$$

Table 3 shows the list of databases currently supported by DBGET/LinkDB in the Japanese GenomeNet service. LinkDB is the database of binary relations containing original links annotated by each database, reverse links and indirect links that are obtained by combining multiple links, and similarity links that are computationally derived by BLAST and FASTA searches. DBGET/LinkDB is a network distributed database system, so that the user can add local databases to be integrated with the GenomeNet databases.

Table 3. The databases supported by DBGET/LinkDB in GenomeNet.

Content	Database	Media
Nucleic acid sequences	GenBank, EMBL	text
Protein sequences	SWISS-PROT, PIR, PRF, PDBSTR	text
3D structures	PDB	text, 3D graphics
Sequence motifs	EPD, TRANSFAC, PROSITE	text
Enzyme reactions	LIGAND/ENZYME	text
Chemical compounds	LIGAND/COMPOUND	text, GIF, 2D graphics
Biochemical pathways	PATHWAY	GIF
Gene catalogs	GENES	text
Amino acid mutations	PMD	text
Amino acid indices	AAindex	text
Genetic diseases	OMIM	text
Literature	Medline (link only), LITDB	text
Link information	LinkDB	text

DBGET/LinkDB allows generic naming of the database names, such as DNA for GenBank + EMBL and protein for SWISS-PROT + PIR + PRF + PDBSTR. This mechanism is used to define the GENES database that is the collection of all gene catalogs in KEGG. The PATHWAY database is the DBGET implementation of the collection of all biochemical pathways in KEGG, where each entry is a GIF image map for the pathway diagram. In addition to GENES and PATHWAY, the LIGAND database [5] is also a product of the KEGG project. LIGAND is a composite database currently consisting of two sections: ENZYME for enzymatic reactions and COMPOUND for chemical compounds.

KEGG system

While the DBGET/LinkDB system is designed to manipulate databases at the entry level, the KEGG system is designed to handle specific data items in each entry of the KEGG databases, such as genes, enzymes, and compounds. The KEGG system is under continuous development, but as of June 1998 it consists of the following programs:

(1) CGI scripts to search and color GIF image maps

(2) Java applets to search, compare, and handle genome maps

(3) CGI scripts to search and handle hierarchical texts

(4) CGI scripts to search and color HTML tables

(5) a suite of software tools for network comparison and pathway reconstruction

The major types of network data shown in Table 2 can be searched in different ways as summarized in Table 4. These capabilities effectively make it possible to perform network comparisons. In fact, it is the main purpose of the KEGG system to treat a network or a subnetwork as a unit. Of course, a KEGG query can be made with a single element against a network, but it is better made with multiple elements against a network to see if, for example, a correct subnetwork (functional unit) is formed.

Table 4. Searching elements in the KEGG network data.

Object	Search by
Pathway map	gene name, EC number, compound number, sequence similarity
Genome map	gene name
Gene/molecule catalog	gene/molecule name, key words
Ortholog group table	sequence similarity

Actually, the traditional sequence alignment is a network comparison problem with relatively stringent conditions; namely, the network is one-dimensional and the order of network elements (amino acids or nucleotides) cannot be changed. The network comparison in KEGG is a more general problem. By comparing a genome and a pathway we wish to identify localized genes in the genome, for example, genes in an operon, that function in a closely related positions in the biochemical pathway. By comparing a sequence similarity network (neighbor) and a pathway we wish to identify duplicated genes forming successive steps in the biochemical pathway. A general algorithm to compare different types of networks was developed (H. Ogata, W. Fujibuchi, and M. Kanehisa, manuscript in preparation), has been used in organizing ortholog group tables (see below), and is provided as software tools in the KEGG system.

Relational database system

An entry of GENES contains, among others, amino acid and nucleotide sequence information that is extracted from the GenBank database, but the description of the gene can be different from GenBank because the functional assignment is re-examined by KEGG. To help this process and to manage updates of the GENES database we use the Sybase relational database system. The system, which is linked to a Web-based gene annotation tool, is limited to internal use.

Data Organization in KEGG

Biochemical pathways

KEGG contains most of the known metabolic pathways, especially for the intermediary metabolism, that are represented by about 100 graphical diagrams (pathway maps). In addition, we are adding various types of regulatory pathways such as membrane transport, signal transduction, cell cycle, transcription, and translation, as well as the information of molecular assemblies. Each pathway diagram is drawn and continuously updated manually. For metabolic pathways the manually drawn diagrams are considered as references of biochemical knowledge containing all chemically identified reaction pathways. The organism-specific pathways are then automatically generated by matching the enzyme genes in the gene catalog with the enzymes on the reference pathway diagrams according to the EC number. The matched enzymes are colored green in the pathway diagrams. This matching process is possible because the intermediary metabolism is relatively well conserved among different organisms. In contrast, the regulatory pathways are too divergent to be represented in a single reference diagram; they are drawn separately for each organism.

Gene catalogs and genome maps

The information of genes and genomes is taken from GenBank and organized as the gene catalog and the genome map. The gene catalog contains classifications of all known genes for each organism. Depending on how one views the function, genes may be classified in different ways. KEGG provides its own classification scheme according to the pathway classification, as well as another scheme by the original authors which is often a variant of Riley's classification [6]. As mentioned the functional assignment of genes is re-examined by KEGG. The genome map is presented to help understand the positional information of genes, such as an operon structure, and its relationship with the pathways and assemblies. Genome maps are manipulated graphically by Java applets.

In order to cope with an increasing number of complete genomes, we are trying to automate as much as possible the EC number assignment that is critical to generate organism-specific metabolic pathways and the gene function assignment. Both assignments are based not only on sequence similarity, but also on additional information including the positional information in the genome and the orthologous relation with different species. Since the operon structure is widespread in bacteria and archaea, the genome map browser has turned out to be an indispensable tool for gene function assignment.

Ortholog group tables

Sequence similarity search against the existing sequence databases often generates a long list of hits, which requires human efforts to screen out orthologous relations that can be used for gene function assignments. The KEGG ortholog group tables are a clean reference data set of orthologous relations that is intended to make this process easier. The table contains not only the information of orthologous and paralogous genes, but also the information of the group of genes that is supposed to form a functional unit, such as a regulatory unit in the metabolic pathway or a molecular unit of assembly. Thus, the ortholog group tables represent a library of 'network motifs' or conserved local network patterns that are related to functional meanings. They are maintained manually with the aid of network comparison tools in KEGG.

Molecular catalogs

The molecular catalogs are to be used for representing functional and structural classifications of proteins, RNAs, other biological macromolecules, small chemical compounds, and their assemblies. However, the current version of KEGG contains only several tables, mostly for enzyme classifications.

Chemical compounds

The living cell contains a number of non-genetic compounds that are synthesized, transported from outside, or simply carried over cell divisions. In order to represent a complete network of molecular interactions, it is necessary to have a complete catalog of compounds in the cell and possibly in the environment as well. The COMPOUND section of the LIGAND database currently consists of over 5,000 chemical compounds, mostly metabolites with links to the location on the metabolic pathways and to the enzymatic reactions involved. The COMPOUND entry also contains the chemical structure that is entered manually using the ISIS system, and the CAS number.

Enzymatic reactions

The information of enzymatic reactions and enzyme molecules is currently stored in the ENZYME section of the LIGAND database. Work is in progress, however, to organize the third REACTION section of the LIGAND database containing both enzymatic and non-enzymatic reactions. A reaction between multi-substrates and multi-products is decomposed into a set of binary relations or approximated by a reaction between two major compounds. The reaction data are especially important for computing possible chemical networks, from which possible gene (enzyme) networks can also be obtained.

Molecular relations

The binary relations of successive enzymes are also extracted from the KEGG metabolic pathway diagrams. They form a class of molecular relation data in KEGG. Another important class of molecular relation is the protein-protein interactions in regulatory pathways such as in signal transduction, cell cycle, and developmental pathways. These data are not yet well organized, except for a few attempts in BRITE

72

(Biomolecular Relation in Information Transmission and Expression) that computerizes the interaction data from literature (see Table 1 for the URL).

Examples of Using KEGG

Navigation

It is recommended to try out the examples described here by opening the KEGG table of contents page (www.genome.ad.jp/kegg/kegg2.html). First of all, this page is organized to represent the network level information of biochemical pathways in the upper portion and the molecular level information of genes and genomes in the lower portion. The user may navigate KEGG resources as well as other integrated resources in a top-down fashion starting from the pathway section or in a bottom-up fashion starting from the gene catalog section. For example, starting from an overall map of the metabolic pathways, the user can zoom into carbohydrate metabolism, and then to citrate cycle (TCA cycle). This is the network level where inter-relationships of molecules are represented in the KEGG pathway diagram as shown in Figure 3. The box is an enzyme with EC number inside, and the circle is a compound with its name written aside. Both are clickable to go down to the molecular level and then to navigate through a number of molecular biology databases.

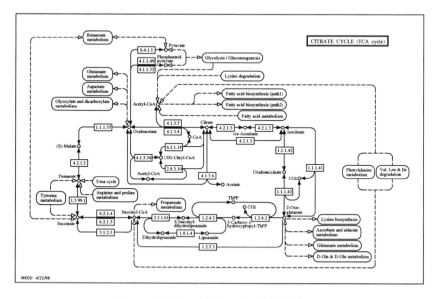

Figure 3: The cytrate cycle in KEGG.

At the upper-left corner of each pathway diagram, there is a pop-up menu with which the user can go to organism-specific pathways and see the differences. For example, *E. coli* has a complete set of genes that forms the TCA cycle, but *H.*

influenzae lacks the upper portion of the cycle shown in Figure 3. Interestingly, *H. pylori* is complementary, having only the upper portion and lacking the lower portion. Furthermore, by examining the KEGG ortholog group table, these two portions turn out to be coded in different operons whenever the operon structure is observed. These observations suggest that the TCA cycle is actually formed by two sets of pathways that are under different regulatory control mechanisms.

Network comparison

The four types of networks (Table 2) can be compared by using the search capabilities of KEGG (Table 4). For example, the genome-pathway comparison is done as follows. Starting from the genome map of a given organism, the user displays the area of interest in the enlarged window and asks where in the known biochemical pathways the genes in the window function. The query can be done by simply clicking on the button marked PATHWAY. A typical result would be a gene cluster in the genome forming a functional unit in the biochemical pathway; namely, the genes in the window code for a set of proteins in successive steps of, say, amino acid biosynthesis.

Another example of network comparison invlolves a hierarchy versus biochemical pathways. For example, in the KEGG table of contents page, select the hierarchical classification (molecular catalog) of enzymes by SCOP 3D folds. By opening the third-level data for beta/alpha (TIM)-barrel in the hierarchy, the user can search all occurrences of TIM-barrel proteins against the known metabolic pathways. This is done by clicking on "Pathway Search by EC" and choosing "Search against 3D structures in PDB" to limit the search for only those enzymes with known structures. One of the results of this query is Phenylalanine, tyrosine and tryptophan biosynthesis, where the last steps of tryptophan biosynthesis are populated by TIM barrel proteins, which suggests possible gene duplications in the evolution of pathway formation [1].

Pathway reconstruction with reference

In the traditional similarity search of individual genes (or proteins) against repositories of all known sequences, it is always problematic to determine an appropriate level of sequence similarity that can be extended to functional similarity. The prediction tools in KEGG incorporate an additional feature that is used for interpretation of sequence similarity; namely, the requirement for reconstructing a complete pathway or a complete functional unit from a set of genes or proteins. The reference for reconstruction is the set of pathway diagrams, and the refined data set of ortholog group tables for a limited, but increasing, number of functional units. For example, by searching sequence similarities against the KEGG ortholog group table for ABC transporters using a set of consecutive genes in the genome as a query (an ABC transporter is often coded in an operon), a transporter can be reconstructed with prediction of substrate specificity according to the subgrouping of the ortholog group table.

The search can also be made against the KEGG pathway diagrams, which form a much larger data set than the ortholog group tables. In this case the search has to be made against a single reference organism, but the procedure is similar; the user specifies a set of sequences as a query. When the EC numbers are preassigned to enzyme genes in the genome, the user can match the set of EC numbers against the KEGG reference metabolic pathways (not the organism-specific pathways). The matched enzymes are marked by color, so that the connectivity and completeness of the marked enzymes can be used to assess the correctness of functional assignment (EC number assignment) in the gene catalog. The existence of a missing element implies either the gene function assignment is wrong or the biochemical knowledge of reaction pathways is incomplete [7].

Pathway reconstruction from binary relations

The prediction above is a homology modeling based on comparison against the well-defined reference. Perhaps, the most challenging task in KEGG is to make predictions even when the reference is missing or incomplete. In the case of the metabolic pathways, if the reconstructed pathway contains a missing element and it is not due to an error in the EC number assignment, then this implies that the reference knowledge is incomplete; an alternative reaction pathway exists or an alternative enzyme with wider specificity takes the place [7]. To investigate this possibility KEGG provides a tool to compute from a given list of enzymes all possible pathways between two compounds and allowing changes in specificity. In our representation, because a list of enzymes is equivalent to a list of substrate-product binary relations, the procedure involves deduction from binary relations, which is like combining multiple links in LinkDB. Possible changes of substrate specificity can be incorporated by considering the hierarchy of EC numbers; namely, allowing a group of enzymes to be incorporated whenever any member of the group is identified in the genome, which effectively increases the number of substrate-product relations. Figure 4 shows the procedure of computing chemical reaction paths in terms of the deductive database.

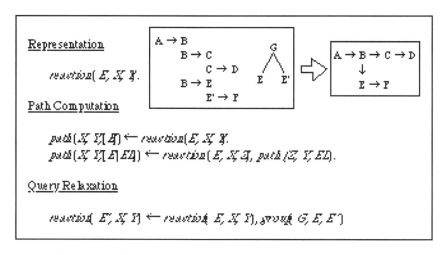

Figure 4. KEGG as a deductive database for path computation.

Future Directions

Computing paths of reasoning

We have mentioned two specific examples of path computation problems: computing multiple links in DBGET/LinkDB and computing chemical reaction pathways in KEGG. This type of deduction from binary relations can further be extended to include different types of relations that are summarized in Table 5.

Table 5. Different types of binary relations

Type	Binary relation
Factual	Cross-references between database entries
Similarity	Computed similarity of sequences or 3D structures
Functional	Functional correlation of interacting molecules
Genomic	Positional correlation of genes in the genome

While factual, similarity, and genomic relations are well computerized, the majority of functional relations, such as pairs of interacting molecules and upstream-downstream relationships in biochemical pathways, exist only in printed materials or experts' brains. In BRITE we have collected a small set of molecular interaction data from primary literature, which is extremely labor intensive and requires expert knowledge. We are also working on to develop computational technologies to automatically extract pairs of interacting molecules from the written text. However, an ultimate solution for fully computerizing interaction data would be direct submission from the authors.

Designing new systematic experiments

The current knowledge of metabolic pathways, especially on the intermediary metabolism, is already well represented in KEGG. The next question is how to organize divergent sets of regulatory pathways. We are collecting pathway data mostly from review articles on various aspects of cellular functions, but the existing literature is the result of the traditional reductionistic approach in molecular biology, which probably represents only a fragmentary portion of actual regulatory pathways in the cell. It is therefore necessary to design new systematic experiments, for example, for protein-protein interactions by yeast two-hybrid systems and for gene-gene interactions by observing gene expression profiles on microarrays. KEGG will provide reference data sets and computational technologies to uncover underlying gene regulatory networks in such experimental data.

Acknowledgments

The KEGG project has been supported by the Human Genome Program of the Ministry of Education, Science, Sports and Culture in Japan.

References

1. Kanehisa, M. *A database for post-genome analysis.* Trends Genet. 13, 375-376 (1997).
2. Kanehisa, M. *Linking databases and organisms: GenomeNet resources in Japan.* Trends Biochem Sci. 22, 442-444 (1997).
3. Goto, S., Bono, H., Ogata, H., Fujibuchi, W., Nishioka, T., Sato, K., and Kanehisa, M. *Organizing and computing metabolic pathway data in terms of binary relations.* Pacific Symp. Biocomputing 2, 175-186 (1996).
4. Fujibuchi, W., Goto, S., Migimatsu, H., Uchiyama, I., Ogiwara, A., Akiyama, Y., and Kanehisa, M. *DBGET/LinkDB: an Integrated Database Retrieval System.* Pacific Symp. Biocomputing 3, 683-694 (1997).
5. Goto, S., Nishioka, T., and Kanehisa, M. *LIGAND: Chemical database for enzyme reactions.* Bioinformatics 14, in press (1998).
6. Riley, M. *Functions of the gene products of Escherichia coli.* Microbiol. Rev. 57, 862-952 (1993).
1. Bono, H., Ogata, H., Goto, S., and Kanehisa, M. *Reconstruction of amino acid biosynthesis pathways from the complete genome sequence.* Genome Research 8, 203-210 (1998)

6 OMIM: ONLINE MENDELIAN INHERITANCE IN MAN

Alan F. Scott*, Joanna Amberger*, Brandon Brylawski** and Victor A. McKusick*

* *Center for Medical Genetics, Johns Hopkins University School of Medicine, Baltimore, MD 21287*

** *National Center for Biotechnology Information, NLM-NIH, Bethesda MD*

Introduction

Mendelian Inheritance in Man (MIM) is a compendium of bibliographic material and observations on inherited disorders and genes maintained by geneticists and molecular biologists. Its online counterpart, OMIM, is freely available on the World Wide Web (WWW). Unlike other databases that maintain primary sequence, mapping, or reference material, OMIM provides authoritative free text overviews about genetic disorders and gene loci that can be used by students, researchers, or practitioners of clinical or molecular genetics. Curation of the database and editorial decisions take place at the Johns Hopkins University School of Medicine. Distribution of OMIM and software development are provided by the National Center for Biotechnology Information at the National Library of Medicine.

MIM was begun by Dr. Victor A. McKusick in the early 1960's. Although available only in print form until the 1980's, it has been maintained on computer since early in its development, largely as an aid to authoring, editing, and publication. In 1987, MIM was made internationally available online as OMIM by the National Library of Medicine, and the current web interface debuted at NCBI in late 1995. MIM is organized by gene locus; however, its emphasis continues to be on content with medical relevance. A detailed history of MIM, its organization, and editorial policies are available in the 12th edition of the book [1] and on the web site (http://www3.ncbi.nlm.nih.gov/omim/).

Figure 1 illustrates the growth of the database in terms of number of entries included in each of the editions of the book. It is currently estimated that this number

78

will plateau at 80,000 to 100,000 when all human genes are known and characterized sufficiently to warrant inclusion.

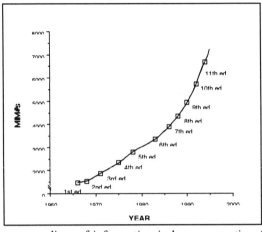

Figure 1. Total number of entries in the printed editions of *Mendelian Inheritance in Man.*

As a comprehensive, authoritative and timely compendium of information in human genetics, OMIM is an important resource for clinicians, teachers, and researchers. For clinicians, OMIM can be used as an aid in differential diagnosis. Physicians and counselors can search the database using key clinical features of a patient. As a teaching tool, OMIM provides students a quick and simple way to find and review essential information about a given gene or genetic disorder. A MEDLINE search for cystic fibrosis, for example, would yield thousands of references. A genetics student would have a difficult time identifying which are the most important. A search of OMIM, however, readily provides this summary information. In research, OMIM serves as a starting point for clinical researchers who want to identify information about genes that may be related to a medical condition (e.g., what mutations are known for a gene and how they manifest themselves) and for basic scientists a search of OMIM may help in identifying disorders that may be caused by a gene that they have characterized.

Perhaps OMIM's greatest utility is to serve as a simple gateway to related information about genetics. Through its wealth of relevant links OMIM serves as an easy access point to databases that contain more detailed information about sequence, map information, references, etc. The ability to easily navigate between databases whose structures and searching schemes are quite different has made OMIM especially popular with users whose need for more detailed data is sporadic and who may not be intimately familiar with other databases and how to use them.

Features of the OMIM database

OMIM entries have a standardized format which includes a number of features, described below, that make it easy, even for the causal user, to obtain information.

The Entry

Each OMIM entry is assigned a unique number. Entries are prepared for each distinct gene or genetic disorder for which sufficient information exists and are not made for genes known only from expressed sequence tags (ESTs), pseudogenes, genetic markers (e.g., "D" segments), or even complete cDNAs for which nothing is known other than the sequence. Many disorders that are not yet characterized at the level of the gene or even at any meaningful biochemical level are included as entries if they show Mendelian inheritance. A major goal of OMIM is to help with the discovery process whereby a gene sequence and a mapped phenotype can be associated. When this occurs the separate entries may be merged.

OMIM authors do not wittingly create more than one entry for each gene locus. The kinds of information that may be included in entries are the approved gene name and symbol (obtained from the HUGO Nomenclature Committee), alternative names and symbols in common use, information about the map location in man and mouse, how it was cloned, information about the protein and DNA sequence such as the size of the gene, the type of product made, what its known or inferred function might be, where the gene is expressed, whether it is related to other similar genes in man or other species and whether there are animal models. For entries where the gene causes a disease, information about key allelic variants is included and clinical details are presented.

Care is taken to assure that distinctive characteristics of given clinical disorders are noted, including variations from the usual case. The information is intended as a guide in diagnosis and management of the particular disorder. Clinical information, given in succinct form, is supplemented by that provided in carefully selected citations that accompany each entry.

Text entries are generally diachronic, meaning that they are added to in chronological order, with the most recent material at the end. This is done, in part, to minimize the effort of having to rewrite thousands of entries each time they are amended and to reflect the historical progression of the knowledge about the locus. Many of the larger entries, for which there is a wealth of information, have been restructured into topical sections. These topics comprise the Table of Contents (see Fig. 2) at the top of the entry. For these entries, new information is added in chronological order into the appropriate section.

User Comments

OMIM encourages users to offer comments about existing entries and suggestions for improvements or additions of materials. Comments relating to scientific and editorial content are forwarded by NCBI to the editorial staff at Johns Hopkins.

Clinical Synopses

Clinical synopses are available for entries that describe a medical phenotype. These single word listings of signs, symptoms, laboratory tests, genetic peculiarities, etc. are written using a controlled vocabulary and provide a quick survey of features of a given disorder. The synopses are particularly useful in creating lists for differential diagnosis.

Allelic Variants

A valuable feature of OMIM are its lists of noteworthy "allelic variants" for a given gene. These are most often disease causing mutations, but can also include common polymorphisms that do not produce disease. In each case, the variants are inherited and are distinct from somatic mutations, as seen in cancer, which are generally not included. OMIM does not try to exhaustively document all known variants at a locus, but rather focuses on those that are relatively common, represent a novel mechanism of mutation, or have historic significance.

The OMIM Gene Map

The OMIM gene map is maintained as a convenience to users and focuses on the "morbid map," i.e., the mapping of disorders. It contains information on chromosomal location, based on linkage and other mapping methods, obtained from the literature cited in the entry. In chromosome-by-chromosome tabular form, the OMIM synopsis of the human gene map gives, for each gene, the chromosomal location, gene symbol, method(s) of mapping, and disorder(s) related to the specific gene. For these mapped disorders, it also indicates whether specific mutations have been identified. The OMIM gene map is not the official map curated by the genetics mapping community and lacks the detail found in maps maintained by GDB and NCBI which are available as links.

Citations

References are highlighted in the text and, if they are cited in PubMed (MEDLINE), the PubMed ID linked to the abstract is listed after the reference. The PubMed links are generated nightly at the time when the database is rebuilt. References are particularly important in OMIM since they are also used to generate links to other databases, including GenBank.

Edit history and credits

When changes are made to an entry by the OMIM editing staff, they are documented in the EDIT HISTORY field with the date and name of the person who made the changes. The CREATION DATE field lists the date that the entry was created and the name of the person who created it. Authors who contribute significantly to the updating or editing of an entry and given credit in the CONTRIBUTORS field.

Searching OMIM

Searching OMIM is as simple as typing in the words of interest. In addition, there are a number of ways to restrict a search (i.e., Boolean operators and field restriction). Entries are retrieved in the order of the number of occurrences of the word in the text. The entries that are retrieved are ranked in order of the number of times the search terms occur in the entry and where it occurs. For example, terms that appear in the title are given greater weight than those occurring in the text or references.

Neighboring

A key new feature of OMIM is the neighboring feature which allows users to perform MEDLINE searches using keywords selected from the text of the preceding paragraph. Each paragraph in the text of an OMIM entry is processed against the entire MEDLINE set of citations using the NEIGHBOR algorithm developed at NCBI. NEIGHBOR finds the MEDLINE articles that seem most closely related to the OMIM paragraph and links them to the paragraph on the OMIM entry's WWW page. This feature permits the user to find MEDLINE references that are germane to his or her specific area of interest in addition to the references in the OMIM entry itself. NEIGHBOR searches are performed against the current literature for the entire OMIM database on a regular basis, thereby assuring that the links from any given entry are up-to-date.

External Links

With the migration of OMIM to the WWW and the proliferation of genetic databases it has been possible to greatly increase the utility of OMIM by providing links to other resources. Among these, as shown in Fig. 1, are links to Entrez, including PubMed, protein and nucleic acid databases, Unigene, the Human Gene Mutation Database, the Genome Database, locus specific mutation databases (such as the CF and PAH Mutation Databases), and the Coriell database of cell lines and probes.

Update log

An update log is available from the home page that takes users to a list of entries that have been changed. The list is arranged by the month and within the month by day. New entries are separated from updated entries; this is particularly helpful for users who only want to browse the database for new entries.

Allied Resources

The home page of OMIM lists a number of allied resources that are distinguished from the external links found in the entries. Paramount among these is the NCBI Entrez MEDLINE and GenBank retrieval system. Resources particularly useful for comparative studies are the databases of the Jackson Laboratory at Bar Harbor, which maintains mouse locus information, the Online Mendelian Inheritance in Animals, and the Seldin/Debry Human/Mouse Homology Map which graphically illustrates corresponding mouse and human homologs and regions of syntenic homology.

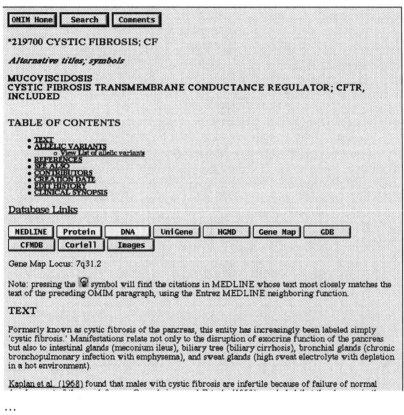

Figure 2: The Cystic Fibrosis entry illustrates several features of the OMIM database. At the top of the page are buttons that allow the user to perform another search, return to the home page or make comments about the entry. The OMIM number is shown along with the approved gene name and symbol. Next are listed alternative names and symbols. The Table of Contents, a feature of the longer entries, takes users to specific regions of the text. Where available, links to other resources are presented. Next, the cytological gene map position is given followed by an explanation of the neighboring feature button. MEDLINE references, including abstracts, are available through the PubMed ID link associated with most references.

How OMIM is Curated

OMIM is maintained by a small team of science writers and editorial assistants. The OMIM director oversees the staff and is aided by deputy scientific directors for "genes" and "phenotypes" who help with the selection of articles from the literature that warrant inclusion. Twelve Subject Author-Editors (SAEs) also help with reviewing the literature and checking entries in their areas of expertise. Genetics graduate students and fellows are employed periodically to review entries and note any obvious factual errors. OMIM periodically invites experts on a given disorder to review and reorganize entries. Only the on-site editorial assistants have privileges to modify the database. This is done to assure, as much as possible, that only people trained in the OMIM authoring software can make direct changes.

OMIM is maintained and edited in a simple text format called OMIM Authoring Format (OAF). OAF is designed to be editable in any text editor and contains only the minimal formatting necessary to permit the OMIM reading software to separate an entry into fields for display and searching. In this way, the authors are not dependent on any specialized software to manipulate the entries and can edit them in a simple, natural way.

OMIM maintains agreements with several major scientific journals to obtain articles a week or so prior to publication. During this time, important, embargoed articles are authored on and entered into the database to be released on the date of publication. In the past year, the database has averaged about 43 new entries and 540 changes in existing entries per month. OMIM does not attempt to be an exhaustive review of the literature. Not only would this be an impossible task given the enormous volume of the world's genetics literature, but it would defeat the useful winnowing process that attempts to select "key" information. Also, the "neighboring" feature of OMIM easily allows the use of the text of a preceding paragraph to search MEDLINE for related articles.

The primary source material for OMIM is the published literature. The director, the deputy directors and the subject author editors review several leading journals that publish major articles in human and molecular genetics. In addition, the director and scientific directors scan the tables of contents of dozens of additional journals for article titles indicating that they may contain information on a new gene or important information to add to an existing gene entry. In the latter case, the articles are photocopied and reviewed as follows. First, a search of OMIM is made to assure that the material is not already entered and whether the new reference constitutes an "add" to an existing entry or warrants a new entry. When DNA or protein sequence is included in a paper, the sequence is compared to GenBank using the NCBI **blast** service [2]. Based on the results of that search, additional papers may be identified that have reported the same sequence, often using a different name. If an approved symbol is not available in GDB then information about the gene is submitted to the HUGO Nomenclature Committee and a request for an approved gene symbol and name is made. The latter are then added to the OMIM entry. The journal article or multiple related articles, supporting sequence comparisons (i.e., **blast** results) and any comments or supporting material are forwarded to a science writer for authoring. OMIM science writers are typically PhDs or MDs with training in genetics. They read the articles and abstract the salient points in the paper. Importantly, the science writers do not simply paraphrase the abstract, but instead add information from the introduction and discussion sections of the citation that attempt to put the work in context and provide sufficient background that a reader can understand the significance of the work.

The Future of OMIM

Perhaps the best measure of a database is how often people use it. At the end of 1995 OMIM received about 4,000 user queries per week. By the middle of 1997, OMIM received over 20,000 searches per week (2,200-2,400 unique users per day). Unlike other databases whose growth will diminish as the human genomic sequence is completed and the map locations of all genes are known with precision, OMIM is likely to continue to grow at an expanded rate. The reason for this is that the windfall of sequence will certainly identify tens of thousands of new genes that will become the fodder for scientists world-wide. As the community reports information about the function and medical significance of these sequences there will be an on-going need to supplement OMIM with that knowledge.

In the near term, OMIM will continue to strengthen its ties to other databases so that there will be a better correspondence of information from one to another. In particular, OMIM will work closely with the HUGO Nomenclature Committee to be certain that approved symbols and gene names are maintained more closely. OMIM will also work with Unigene to assure that all OMIM entries are represented in the Unigene database and that well documented genes from Unigene are included in OMIM. A challenge to OMIM will be to maintain standards of authoritativeness, thoroughness and timeliness as it deals with the increasing torrent of new genetic discoveries. Ultimately we hope that OMIM will be seen as a Rosetta stone for the genetics community allowing scientists, students and clinicians to easily obtain relevant information from all pertinent sources on all human genes.

References

1. McKusick, V.A. *Mendelian Inheritance in Man. Catalogs of Human Genes and Genetic Disorders.* Baltimore: Johns Hopkins University Press, 1997 (12th edition).
2. Altschul, S.F., Gish, W., Miller, W., Myers, E.W. and D.J. Lipman, J. Mol. Biol. 215, 403-10 (1990).

7   GDB: INTEGRATING GENOMIC MAPS

Stanley Letovsky

Cereon Genomics, LLC; 45 Sydney St., Cambridge, MA 02139

Introduction

The Genome Database[1] (GDB) [1] is a public database of human genome mapping data. The main classes of data in GDB are genomic maps, along with the objects located on the maps, such as genes, clones and markers (STSs), and marker polymorphisms. GDB does not contain human sequence data, though it does contain links from its objects to entries in the sequence databases. GDB can be viewed as a species-specific genome database for the human, so it attempted to play a role analogous to that played by MGD for the mouse, flybase for drosophila, and ACEDB for C.elegans.

GDB is implemented using a relational database management system with an object-oriented layer on top, based on OPM [2]. The primary user interface is the Worldwide Web, including a Java map viewer. Many of the Web interface components are generated automatically from the object-oriented schema by the Genera Web/database gateway [3]. The system supports direct community curation of its contents by authorized users; there is also an electronic data submission format for file-based submission.

In its earliest years GDB's representation of maps included a coordinate-based representation for cytogenetic data, as well as a textual representation for other sorts of mapping data. Both representations were limited in their ability to support searches or graphic displays. Subsequently the representation of maps was consolidated and generalized to cover the richer range of maps that were increasingly being generated in the 90's, including linkage, radiation hybrid and STS-content maps. Query and display tools were implemented based on the newer representations that allow selected portions of maps to be retrieved and graphically displayed.

---

[1] Funded until recently by the US Department of Energy, National Institutes of Health and Japan's Science and Technology Agency.

A common problem in the construction of databases of mapping information is how to optimally align or integrate mapping data from different methods, such as genetic or physical mapping, or different sources, such as radiation hybrid maps produced by different labs or from different mapping panels. There main motivations for aligning maps are to support database searches of chromosomal regions of interest and to produce better graphic displays. Several approaches to these problems have been explored and implemented within GDB; this article describes and critiques those methods.

Map Querying

An important query for GDB is to find all loci in a region of interest, sometimes with additional restrictions as to the type of locus, existence of polymorphisms, or functional category. The region of interest might in principle be specified in any of several ways, such as the region between two specified loci, or the neighborhood of a specified locus for some number of units in each direction. GDB stores many maps of many different types, and it is desirable to have such positional search across all maps of a region simultaneously. Intuitively, we want the database to function as a stack of aligned maps, and a query to cut a thick slice through that stack (see Figure 1). A central concern of this article is how to best align the maps for this purpose.

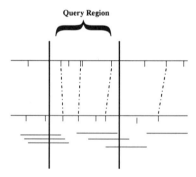

Figure 1: Query cutting through a stack of aligned maps

The details of the relational implementation of overlap queries are worth pointing out briefly. Each locus is considered to have a localization interval, consisting of a minimum and a maximum coordinate[2]. The query to retrieve stored intervals I overlapping a query interval Q can be expressed intuitively as the negation of nonoverlap - *find intervals I which are not disjoint from Q. I* can be disjoint from Q

---

[2] These follow naturally from binned maps, where the coordinates are those of the bin boundaries; for cytogenetic maps they are the coordinates of the band boundaries, and in general they are coordinates of backbone markers. Backbone or point-like markers are represented by zero-width intervals. Distance-based linkages can be converted to intervals by using a suitable multiplier of the standard error.

in one of two ways: by being wholly to the left of it, or wholly to the right. The query is thus:

$$\text{not}(I_{max} < Q_{min} \text{ OR } Q_{max} < I_{min})$$

Q_{min} and Q_{max} are constants in the query, whereas I_{min} and I_{max} are data columns. Negation and disjunction (OR) operations can confuse relational query optimizers, so queries written in this form will typically not execute efficiently. An equivalent query which can usually be executed efficiently is:

$$I_{max} >= Q_{min} \text{ AND } Q_{max} >= I_{min}$$

This query is especially efficient if a two-column index is placed on (I_{min}, I_{max}). This allows searches to be restricted to the region of interest Q; without that index queries may scan from Q_{min} to the maximum end of the chromosome, and intersect the results with the results of scanning from Q_{max} to the minimum end of the chromosome, which means the entire chromosome is scanned for every query; this is much slower.

Note also that the more restrictive query to find stored intervals I contained in query interval Q has the form
$$Q_{min} <= I_{min} \text{ AND } I_{max} <= Q_{max}$$
which is efficient under the same indexing as the overlap query.

Linear Alignment

In order to search a set of aligned maps we must provide each locus with a coordinate in a common coordinate system, which I will call *a universal coordinate system*. In general each map may have its own coordinate system, involving different origins and units (centiRays, centiMorgans, kilobases, order-only). To define universal coordinates we choose some map as the standard map for each chromosome, and align all other maps to that map. The choice of standard map is made by human judgement, paying attention to criteria such as marker density, accuracy, and use of markers shared by other maps.

Our first implementation of a universal coordinate system used linear alignments. For each map of the chromosome a linear regression was performed on the coordinates of the markers that the map had in common with the chosen standard map, using the nonstandard map coordinates as the independent (X) variable and the standard map as the dependent (Y) variable. Figure 2 shows an example; the outliers are due to markers placed in very different positions on the two maps. The resulting regression line (blue) is used to map the coordinates of all loci in the nonstandard map into the universal coordinates, i.e., the coordinates of the standard map. Once universal coordinates have been assigned to all map elements on all maps, searches against these coordinates can retrieve loci regardless of what primary map they belong to.

88

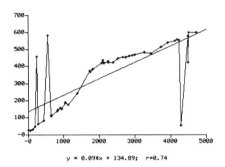

y = 0.09%x + 134.89; r=0.74

Figure 2: Whitehead Radiation Hybrid Map vs. Stanford RH Map of Chr.3

(The nonlinear relationship evident between the two maps is somewhat surprising; this is to be expected between linkage and physical maps because recombination is inhibited near the centromeres and telomeres, resulting in a compression of genetic distances in those areas and a relative expansion in the chromosome arms, where most recombination takes place. In the figure both maps are radiation hybrid maps, which might have been expected to be more linearly correlated. The nonlinearity may be a function of the different radiation doses used to create the panels the maps were based on, which affects the density of breakpoints. If the centromere is nonrandomly retained in radiation hybrid cells, then the different panels might experience different degrees of centromeric distortion. However these are just guesses; to my knowledge this effect has not been well studied.)

Although linear transformation provides a practical solution to the map alignment problem, it is by no means perfect. With any universal coordinate scheme a locus which appears on several maps may be assigned slightly different universal coordinates in each of them, and so will appear several times in several different places in the universal coordinate space. Ideally these places should be close to each other, unless the locus was incorrectly placed in one of the maps, or a locus identification error was made during entry of the map into GDB. Figure 3 shows one way to visualize this effect, which we call a dispersion plot: each x,y point represents a locus having a universal coordinate x from one map, and y from another. (The same pair of maps will also generate a point for this locus at y,x, so the plot is symmetric.) In the ideal case where every locus ended up at a unique universal coordinate all the points would fall along the line $y=x$. This distance of points from this line is a measure of the variability in the assignment of universal coordinates, and the overall thickness of the distribution intuitively represents how good the linear transformation approach is.

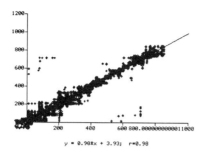

$$y = 0.98*x + 3.93; \quad r=0.98$$

Figure 3: Universal Coordinate Dispersion Plot of Chromosome 3

How important is this dispersion? Is it worth worrying about? When universal coordinates are used to support queries of the content of a region, dispersion can cause both false negative and false positive errors in the results, i.e., they can fail to retrieve the appropriate parts of some maps and retrieve inappropriate parts of others (see Figure 4). In figure 3 the dispersion appears to be on the order of about 5% of the chromosome length; intuitively that means that at resolutions finer than this the alignments are not reliable.

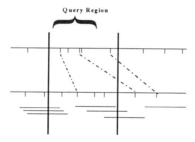

Figure 4: Query region does not find completely corresponding regions of maps due to nonlinear alignment

One surprising aspect of dispersion is that there is no dispersion associated with points that occur on only one map, because a single regression line transforms them to a unique universal coordinate. It happens that cytogenetic band boundaries occur only on the cytogenetic maps, so they get assigned a single unique coordinate, as if they were very precisely localized. A marker which has been used as a landmark on many maps will by contrast be associated with a cloud of points in universal coordinates, suggesting that it has a fuzzy localization, whereas in fact we have much more information about its position than we do about something like a band boundary. When cytogenetic bands are used to define the region of interest for a positional query, this spurious accuracy in their universal coordinates can give anomalous results: regions of non-cytogenetic maps which are known to be located in a given band on the cytogenetic map may not be retrieved by a universal coordinate query on that band. As the maps in GDB increasingly come to span multiple levels of resolution all the way down to sequence features, errors such as these become more serious.

Nonlinear Alignment

There are a number of reasons why points in the dispersion plot do not fall on the line $y=x$. These can include measurement error in mapping, mistakes made in mapping or data entry, as in the outliers in figure 2, which reflect dramatic disagreements between two maps as to the position of a locus. More importantly, distances in one map may have a nonlinear relationship to distances in another. For example, genetic and physical distances typically have a non-linear relationship because recombination occurs more frequently near the middle of the chromosomal arms, causing a high ratio of genetic to physical distance in those regions, but a low ratio in the pericentromeric regions where recombination is inhibited. This nonlinear relationship between different types of map distances means that the regression line will be a more accurate transformation in some regions than in others.

One solution to the problem of nonlinearity is to use a nonlinear transformation function which warps the maps into better correspondence. The best transformation function would be the curve defined by the common markers, which we can approximate by a piece-wise linear function. However the transformation must be monotonic (nondecreasing) if it is to preserve marker orders when transforming a map into universal coordinates, and we have seen how order discrepancies between maps can introduce outliers in the plots. To remove these we select a maximal subset of the common markers that are order-consistent between the two maps. Such a subset is called *a longest monotonic chain*. A piecewise-linear function is then defined over that set of points. Figure 5 shows the dispersion for a set of maps aligned with the longest chain method as compared with linear regression. The reduction in dispersion relative to the linear transformation is apparent.

Special care must be taken at the ends of maps, where there may be loci that occur beyond the ends of the piecewise linear function. The function must be extrapolated to handle such points. Two obvious solutions are to use the regression line slope or the slope of the last linear segment; we have found the former to be more robust.

Although longest chain piecewise linear alignment is clearly an improvement over linear regression, it can probably be further improved on. The algorithm chooses arbitrarily among equally long longest chains; it should be possible to somehow average among all possible longest chains. Also it might be worth considering other measures of length besides number of points, such as distance covered, or composite measures that include both number of points and distance spanned

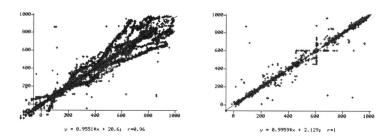

$y = 0.9551*x + 20.6;$ $r=0.96$ $y = 0.9959*x + 2.129;$ $r=1$

Figure 5: Dispersion of linear (left) and longest chain (right) universal coordinates for chromosome 1

Measuring Dispersion

Dispersion can be measured in a number of ways, of which the most straightforward is perhaps the standard deviation of the distance of all the points from the $y=x$ line; this number can be interpreted as a distance error in universal coordinates. A somewhat better measurement would compare each universal coordinate for a locus with the standard-error-weighted average of the universal coordinates for that locus; this would have the desirable property that a locus with N coordinates would contribute only N values to the overall dispersion; the pairwise comparisons shown here generate N*(N-1)/2 values, which means that common loci are overrepresented in the results. Dispersion measures can be computed for individual maps, by comparing the universal coordinates of loci on that map with those from all the other maps; the resulting map dispersion number can be used to rank maps by their overall concordance with other maps, and can also be interpreted as a standard error for the universal coordinates of that map.

Similarly, dispersion can be computed on a per-locus basis to identify highly discrepant loci. The scripts which compute universal coordinates for GDB generate HTML reports showing map-dispersion (figure 6) and the most discrepant loci (figure 7)

Map Details

Map	Std Err	N
Whitehead Contig WC1.19 (Nov 1996)	0	1
Whitehead Contig WC1.9 (Nov 1996)	0	5
Whitehead Contig WC1.4 (Nov 1996)	0	1
Whitehead Contig WC1.1 (Nov 1996)	0	1
Whitehead Contig WC1.6 (Nov 1996)	0	1
Whitehead Contig WC1.5 (Nov 1996)	1	10
Whitehead Contig WC1.8 (Nov 1996)	2	6
Whitehead Contig WC1.3 (Nov 1996)	3	9
Whitehead Contig WC1.7 (Nov 1996)	4	39
Whitehead Contig WC1.16 (Nov 1996)	4	41
Whitehead Contig WC1.23 (Nov 1996)	4	69
Whitehead Contig WC1.22 (Nov 1996)	5	19
Whitehead Contig WC1.13 (Nov 1996)	5	4
Whitehead Contig WC1.10 (Nov 1996)	6	25
Whitehead Contig WC1.12 (Nov 1996)	6	12
1q41-43 Content Contig	6	65
Whitehead Contig WC1.11 (Nov 1996)	8	23
Whitehead Contig WC1.2 (Nov 1996)	9	4
Whitehead Contig WC1.21 (Nov 1996)	9	29
CHLC Chromosome 1 Recombination Minimization (Sex Ave)	9	463
CHLC Chromosome 1 Recombination Minimization (Female)	9	437
CEPH/Genethon Chromosome 1 Linkage Map	13	752
Whitehead Contig WC1.14 (Nov 1996)	14	299
CHLC Chromosome 1 Recombination Minimization (Male)	14	354
Genethon - Chromosome 1 (March 1996)	15	1030
SHGC Chromosome 1 Radiation Hybrid Map (G3)	15	765
Whitehead Contig WC1.15 (Nov 1996)	19	119
Whitehead DR11 (Nov 1996) - Chromosome 1	30	1575
RH Consortium Transcript Map - Chromosome 1	37	1782
Whitehead Contig WC1.17 (Nov 1996)	45	502
Chromosome - 1 - Cytogenetic Map	68	83
Whitehead Contig WC1.20 (Nov 1996)	80	209

Figure 6: Maps ordered by increasing dispersion measure

Most Discrepant Loci

Locus	AvgPos	StdErr	Map	MapPos
SHGC-30207	478	272	RH Consortium Transcript Map – Chromosome 7	94
			Whitehead DR11 (Nov 1996) – Chromosome 7	671
			RH Consortium Transcript Map – Chromosome 7	670
SHGC-10096	340	255	RH Consortium Transcript Map – Chromosome 7	84
			Whitehead Contig WC7.10 (Nov 1996)	595
SHGC-9965	348	255	RH Consortium Transcript Map – Chromosome 7	94
			Whitehead Contig WC7.8 (Nov 1996)	603
SHGC-31100	386	167	RH Consortium Transcript Map – Chromosome 7	151
			Whitehead DR11 (Nov 1996) – Chromosome 7	509
			RH Consortium Transcript Map – Chromosome 7	500

Figure 7: Loci with large dispersion

The use of dispersion as a metric to evaluate the quality of alignments suggests the possibility of using dispersion minimization to drive the alignments. A simple way to do this would be to choose a fixed set of control points for each nonstandard map (the x_i's in figure 8) and choose y_i's that determine a piecewise-linear transformation which minimizes dispersion. The optimization problem is linear and has a small number of variables, so it should be easily solvable.

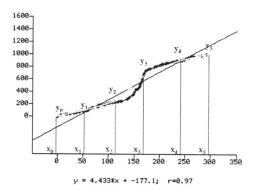

$$y = 4.433*x + -177.1; \quad r=0.97$$

Figure 8: Direct minimization of dispersion

Map Integration

Another approach to the query problem is to integrate the maps, and use coordinates in the integrated map as universal coordinates for querying purposes. What does integration mean? Don't the universal coordinate procedures described above perform a sort of integration, by placing all the elements of all the maps in a common framework? Yes and no. The universal coordinate approach can be viewed as producing a universal map which has one element for each map element in each of the source maps. A marker that appears in several maps will be associated with multiple map elements in such a universal map. An integrated map, by contrast, would put every marker in just one position, no mater how many primary maps it occurred in. That position should have an associated window of uncertainty, and should be influenced by the marker's positions in the different source maps. The uncertainty should be truly reflective of the available information on the marker's position. Integration should ideally also eliminate the need to arbitrarily select one map as a standard, which introduces biases. The coordinates and uncertainties resulting from a meaningful integration along these lines might yield significantly improved query behavior over universal coordinates. See [4,5] for work in this area. Unfortunately, meangful integrated maps are considerably more difficult to construct than alignments; a key issue is how to combine order data from multiple maps and preserve it in the integrated map. There are also a number of technical difficulties associated with defining a meaningful notion of an uncertainty window for a locus on an integrated map. Mapping uncertainty is normally associated with pairwise distances, yet for a map to be efficiently searchable it must use coordinates rather than intermarker distances. The standard error of a coordinate assignment as not a well defined concept, and yet some such notion is necessary to implement an appropriately fuzzy search. This is an important area for future work that affects the positional querying semantics, the integration algorithm, and the graphic display of integrated maps.

Map Display

The MapView 2.0 Java™ applet currently used in GDB can display multiple maps with alignment lines between common markers (figure 9).

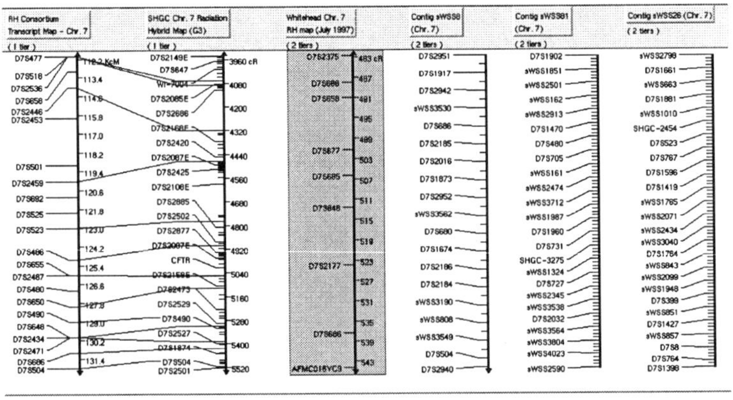

Figure 9: MapView 2.0 shows the same region in several maps.

One problem with such displays is that the number of maps keeps increasing. After doing a positional query the user is currently asked to select which of the maps intersecting the query region they wish to see in the viewer. That list keeps getting longer, and the user often has no intelligent basis for making the choice. We would like to provide a robust default for this choice: an integrated map whose content synthesizes that of all the other maps.

Integrated maps can be constructed based on piecewise-linear alignments. To produce these maps the various universal coordinate localizations for a locus must be combined into a single interval; at GDB we use a modified form of unioning which throws out poor localizations that are consistent with tighter ones. The resulting integrated or *comprehensive* map (figure 10) can be useful for certain purposes, but must be treated with some skepticism. Anyone wanting detailed order information in particular is well advised to look at the original maps, not the integrated one.

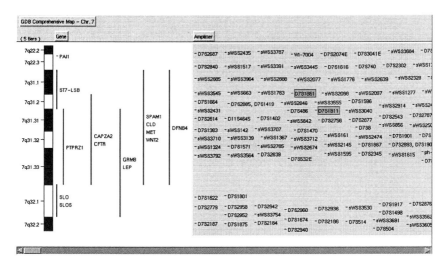

Figure 10: A piece of an "comprehensive map" derived from aligning maps and collapsing multiple localizations by unioning. The width of the localization intervals and the apparent orders cannot be trusted at high resolutions.

As long as precision is not crucial, comprehensive map coordinates can be useful for graphically displaying the results of locus queries, as shown in the following figures..

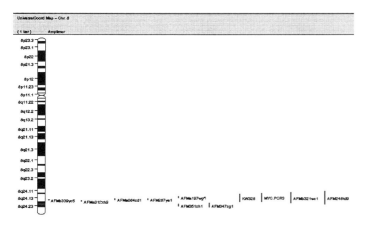

Figure 11: graphic display of results of a query for polymorphic markers in a region of interest

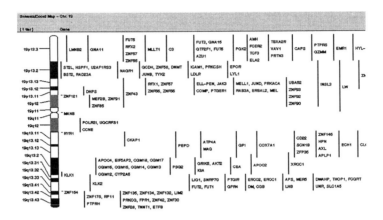

Figure 12: Comprehensive map of genes on chromosome 19

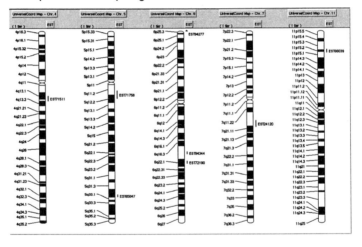

Figure 13: Comprehensive map of EST's expressed in fetal brain, showing results on multiple chromosomes

Discussion

Static alignment of multiple maps of a genome can be useful to support queries by region of interest. Efficient implementation of such queries in relational databases is possible. Nonlinear alignments are significantly better than linear ones for querying purposes. Integrated maps derived from nonlinear alignments can be useful for visualization if used with appropriate caution.

It is worth noting that other approaches to this problem are possible. Dynamic alignment aligns maps at query time about the markers of interest; this strategy is used by Chromoscope in the Genomes Division at NCBI [6]. The resulting alignments are better for the specific marker of interest; however it is not clear that this strategy is applicable to the problem of querying a relational database for a region of interest.

An important area for future work is the extension of these ideas to comparative maps across multiple species. Within GDB some initial work was done aligning mouse maps to human maps within the universal coordinate framework: the mouse maps were rearranged to optimize the alignment with human. This allowed mouse genes to be retrieved in human regions of interest, which can be a useful mechanism for suggesting candidate gene locations. An issue for the future is the design of a more general system that allows mapping information to flow across species.

References

1. GDB, www.gdb.org.
2. Markowitz et al, OPM: see article in this collection.
3. Letovsky, S. Genera: http://cgsc.biology.yale.edu/genera.
4. Letovsky, S. and Berlyn, M. *CPROP: A Rule-Based Program for Constructing Genetic Maps*. Genomics 12:3 pp.435-446.
5. Collins A., Frezal J., Teague J. & Morton NE. (1996). *A metric map of humans:23,500 loci in 850 bands*.Proc. Natl. Acad. See also LDB, http://cedar.genetics.soton.ac.uk/public_html.
6. Entrez Genomes Division, http://www.ncbi.nlm.nih.gov/Entrez/Genome/org.htm

8   HGMD: THE HUMAN GENE MUTATION DATABASE

Michael Krawczak, Edward V. Ball, Peter Stenson and David N. Cooper

Institute of Medical Genetics, University of Wales College of Medicine
Heath Park, Cardiff CF4 4XN, UK
Phone: (+44) 1222 744062 (DNC) or (+44) 1222 744957 (MK)
Fax: (+44) 1222 747603
Email: cooperdn@cardiff.ac.uk or krawczak@cardiff.ac.uk

Summary

The *Human Gene Mutation Database (HGMD)* represents a comprehensive core collection of data on published germline mutations in nuclear genes underlying human inherited disease. By June 1998, the database contained in excess of 13,500 different lesions in a total of 766 different genes, with new entries currently accumulating at a rate of over 2,500 per annum. Although originally established for the scientific study of mutational mechanisms in human genes, *HGMD* has acquired a much broader utility to researchers, physicians and genetic counsellors so that it was made publicly available via the internet at *http://uwcm.ac.uk/uwcm/mg/hgmd0.html* in April 1996. Mutation data in *HGMD* are accessible on the basis of every gene being allocated one webpage per mutation type, if data of that type are present. Meaningful integration with phenotypic, structural and mapping information has been accomplished through bi-directional links between *HGMD* and both the *Genome Database (GDB)* and *Online Mendelian Inheritance in Man (OMIM)*, Baltimore, USA. Hypertext links have also been established to Medline abstracts through *Entrez,* and to a collection of 584 reference cDNA sequences used for data checking. Being both comprehensive and fully integrated into the existing bioinformatics structures relevant to human genetics, *HGMD* has established itself as the central core database of inherited human gene mutations.

Introduction

The *Human Gene Mutation Database* (*HGMD*), maintained at the Institute of Medical Genetics in Cardiff, represents a comprehensive core collection of data on germline mutations underlying human inherited disease. Thus, *HGMD* comprises published single base-pair substitutions in coding, regulatory and splicing-relevant regions of human nuclear genes as well as deletions, duplications, insertions, repeat expansions and "indels", plus a number of complex gene alterations and rearrangements not covered by the above categories. Somatic gene mutations and mitochondrial genome mutations are not included.

The curators of *HGMD* have adopted a policy of entering each mutation only once in order to avoid confusion between recurrent and identical-by-descent lesions. Reliable discrimination between these two alternatives would require information available only for a very small proportion of known lesions. Therefore, although data on the regional, ethnic and haplotype context of mutations would be extremely useful in terms of epidemiological and population genetics research, any unselective accumulation of literature reports would have resulted in an inflation of references with little immediate scientific use.

Although originally established for the scientific study of mutational mechanisms in human genes (1), *HGMD* has acquired a much broader utility in that it provides information of practical importance to researchers in human molecular genetics, physicians interested in a particular inherited condition in a given patient or family, and genetic counsellors. In view of its potential usefulness, the curators of *HGMD* made the database publicly available (2) through the WorldWideWeb in April 1996.

Data coverage and structure

By June 1998, *HGMD* contained over 13600 different lesions in a total of 766 different genes (Table 1). Entries are currently accumulating at a rate of over 2,500 per annum. Coverage is limited to original published reports although some data are taken from "Mutation Updates" or review articles. Mutations reported only in abstract form are not generally included. Data acquisition for *HGMD* has been accomplished by a combination of manual and computerised search procedures, scanning in excess of 250 journals on a weekly/monthly basis.

Table 1. Number of *HGMD* entries by mutation type (June 1998)

Mutation Type	No. of entries
Single base-pair substitutions, missense/nonsense	8495
Single base-pair substitutions, splicing	1308
Single base-pair substitutions, regulatory	118
Small deletions (≤20bp)	2245
Small insertions (≤20bp)	807
Small indels (≤20bp)	106
Repeat expansions	16
Gross deletions (>20bp)	370
Gross insertions and duplications (>20bp)	81
Complex rearrangements including inversions	96
Total	13642

All *HGMD* entries comprise a reference to the first literature report of a mutation, the associated disease state as specified in that report, the gene name, symbol (as recommended by *HUGO*) and chromosomal location. In cases where a gene symbol has not yet been made available owing to the recency of the cloning report, a provisional symbol has been adopted which is denoted by lower-case letters. Single base-pair substitutions in coding regions are presented in terms of a triplet change with an additional flanking base included if the mutated base lies in either the first or third position in the triplet. While substitutions causing regulatory abnormalities are logged in with eight nucleotides flanking the site of mutation on both sides, no flanking sequence has been included yet for substitutions leading to aberrant splicing. Micro-deletions and micro-insertions (of less than 20 bp) are presented in terms of the deleted/inserted bases in lower case plus (in upper case) 10 bp DNA sequence flanking both ends of the lesion. Either the codon number or, in cases where a lesion extends outwith the coding region of the gene in question, other positional information, is provided e.g. 5' UTR (5' untranslated region) or E6I6 (denotes exon 6/intron 6 boundary). Codon numbering may in some cases display inconsistencies owing to the common use of different numbering systems for the same protein. For the majority of genes, however, residue numbering has been standardized with respect to a generally accepted numbering system employing the appropriate reference cDNA sequence. For gross deletions, gross insertions and complex rearrangements, information regarding the nature and location of a lesion is logged in narrative form because of the extremely variable quality of the original data reported.

Summary data on the frequencies of different types of amino acid substitutions, the location of splicing relevant single base-pair substitutions and the sizes of micro-deletions and insertions are presented in table form. Mutation maps of every gene provide a graphical display of mutations within the coding regions so that the distribution of such lesions can be assessed at a glance.

Data access

HGMD is accessible on the basis of every gene being allocated one webpage per mutation type, if data of that type are present. Since *HGMD* is partly dependent upon industrial funding and considerable editorial work is involved in data curation over and above mere literature screening (e.g. to ensure the consistency of nucleotide sequence information, amino acid residue numbering and gene symbol usage), unsolved copyright problems have so far precluded *HGMD* from being downloadable in its entirety. However, once the closer cooperation with publically funded bioinformatics institutions currently envisaged has been put in place, unrestricted access to the database will become possible. *HGMD* currently serves in excess of 200,000 search requests per annum.

Meaningful integration of the data with phenotypic, structural and mapping information on human genes has been accomplished through bi-directional links between *HGMD* and both the *Genome Database (GDB)* and *Online Mendelian Inheritance in Man (OMIM)*, Baltimore, USA. In addition, hypertext links have been established from *HGMD* references to Medline abstracts through *Entrez*. Hypertext links have also been set up to "reference cDNA sequences" (584 to date) which are also used for data checking. "Genomic reference DNA sequences" are currently being constructed which will provide data on exon-intron junctions and the location (if known) of the transcriptional initiation sites.

The links to *GDB* and *OMIM* have enforced the standardisation of disease and gene nomenclature in *HGMD*. Thus *HGMD* can be searched either by *HUGO*-approved gene symbols, *GDB* accession numbers, or *OMIM*-compatible disease or gene names. For genes for which Locus-Specific Mutation Databases are available on the Internet, these databases (currently ~70) can be accessed either from the corresponding gene-specific *HGMD* pages or via the Locus-Specific Mutation Database page (3).

Conclusions and Outlook

Being both comprehensive and fully integrated into the existing bioinformatics structures relevant to human genetics, *HGMD* has established itself as the central core database of inherited human gene mutations. Looking to the future, efforts will be made to improve the provision of flanking sequence data, to increase the number of cDNA and genomic reference sequences provided and to make the data collections on gross gene lesions and disease-relevant polymorphisms fully comprehensive.

In order to improve the accuracy, efficiency and rapidity of mutation publication, however, direct submission of mutation data to a central resource capable of (and responsible for) checking the novelty and consistency of data is both necessary and desirable. Although some Locus-Specific Databases have included mutations not published anywhere in the literature, even the close integration of these facilities will be inadequate to the task of meeting the demands likely to be made upon a central data repository. A substantial proportion of published mutation data are derived from genes in which only a handful of lesions have so far been characterized (Table 2); such genes do not therefore warrant the establishment of a Locus-Specific Database. In part for this reason, Locus-Specific Databases are currently accessible via the Internet for only ~9% of genes referred to in *HGMD*. Although mutation data associated with these genes should comprise ~48% mutations in *HGMD* (assuming the Locus-Specific Databases to be sufficiently comprehensive), the obvious lack of general coverage emphasizes the point that comprehensive collection of mutation data can only be performed in generalised fashion. To this end, *HGMD* has instituted a collaboration with Springer-Verlag GmbH, Heidelberg, to make online submission and electronic publication of human gene mutation data possible (4). These data are being published regularly by Springer's journal *Human Genetics* in both electronic and printed form. Once published, the data will be transmitted to Cardiff to be deposited in *HGMD*.

Table 2. Number of *HGMD* entries per gene by mutation type (June 1998)

Number of entries per gene

Mutation type	1	2	3	4-5	6-10	11-25	26-50	51-100	>100
Single bp substitutions									
Missense/nonsense	154	105	49	65	89	107	47	28	8
Splicing	113	69	31	26	22	26	5	2	0
Regulatory	30	5	2	3	3	2	0	0	0
Other lesions									
Small deletions (≤20bp)	121	69	35	52	35	30	14	4	1
Small insertions (≤20bp)	114	48	20	21	23	14	2	0	0
Small indels (≤20bp)	50	18	1	1	2	0	0	0	0
Repeat expansions	16	0	0	0	0	0	0	0	0

References

1. Cooper DN, Krawczak M (1993) *Human Gene Mutation*. BIOS, Oxford.
2. http://www.uwcm.ac.uk/uwcm/mg/hgmd0.html
3. http://www.uwcm.ac.uk/uwcm/mg/oth_mut.html
4. http://link.springer.de/journals/humangen/mutation

Acknowledgements

The authors wish to thank SmithKline Beecham, Pfizer, the Genome Database and the Deutsche Forschungsgemeinschaft for their financial support and Iain Fenton for computer assistance.

9 SENSELAB: MODELING HETEROGENOUS DATA ON THE NERVOUS SYSTEM

Prakash Nadkarni[1], Jason Mirsky[2], Emmanouil Skoufos[1,2], Matthew Healy[3], Michael Hines[2], Perry Miller[1] and Gordon Shepherd[2]

[1]Center for Medical Informatics, Yale University School of Medicine, New Haven, CT
[2]Department of Neurobiology, Yale University School of Medicine, New Haven, CT
[3]Research Division, Bristol-Myers-Squibb Pharmaceutical Corporation, Wallingford, CT

Introduction

Knowledge about the nervous systems (NS) of higher organisms is evolving rapidly, and databases that store this information will become valuable research resources. The growth of knowledge is occurring in multiple dimensions, with advances at the molecular/sequence level matched by discoveries in gene / gene product influence and interaction. The molecular/functional data must be correlated with corresponding data at the gross anatomical or pharmacological level, such as neurotransmitter/receptor distribution, or the locations where the axons of particular neurons project.

This proliferation of knowledge poses interesting informatics challenges. During the course of the SENSELAB project (1, 2), which has been funded through the Human Brain Project (3)), our group has dealt with the problem of representing several kinds of data, through creation of multiple, physically independent databases. We briefly summarize this work to indicate the nature and scope of the data involved.

- **A Web-accessible database (ORDB) of olfactory receptor sequences**. Both amino acid and nucleotide sequences are represented, and some sequences have associated 3-D structural information in PDB format.This work, previously described in **(4)**, was done by Matthew Healy, Michael Singer, Jason Smith and Emmanouil Skoufos. The need for ORDB was suggested by Doron Lancet of the Weizmann Institute, Israel.

- **A database of computational models of neuronal function, (ModelDB)** (5) that facilitates the creation and running of neuronal models over the Internet through a Web interface. The initial work, using the GENESIS neuronal simulator (6) was done by Bret Peterson. Subsequent work, described in (7) and done primarily by Jason Mirsky and Michael Hines, permits the use of an alternative simulator, NEURON (8).

- **NeuronDB** (9), **a database of neuronal types, with associated receptors, neurotransmitters, canonical compartments, ion channels and relevant literature citations.** Such data is multi-axial (multi-dimensional) and NeuronDB permits a user to query the data from any axis (e.g., a receptor type), and return associated information on other axes (e.g., channel type). This Web-accessible system, primarily implemented by Jason Mirsky and Gordon Shepherd, mostly contains data on neurons of the olfactory system.

- **OdorDB, a database of odor molecules and associated experimental results and literature citations** (10). Structural formulae (as graphics) are among the data associated with each molecule, with links to NeuronDB, ORDB and NCBI's PubMed.. This work has been done primarily by Emmanouil Skoufos and Prakash Nadkarni.

Currently, these databases are implemented with different database engines. (E.g., NeuronDB and ModelDB use the Illustra Object-Relational Database engine.) The databases communicate with each other, when necessary, through Web hyperlinks (query strings passed through URLs to CGI scripts).

We are now considering tighter coupling of these systems. The most obvious way to achieve this is to combine the data into a single physical database, while preserving the look and feel of the existing Web front ends. Physical integration significantly simplifies certain tasks such as creating queries that bridge the (presently separate) components. One consideration for successful integration, is whether the schema of the integrated system should be simply a merge of the individual schemas or a complete redesign. While the former approach is straightforward, the latter approach touches on several interesting informatics research issues, which are the focus of this article.

The Challenge of Managing Highly Heterogenous Data

While the data in SENSELAB is varied, it currently falls short of the entire potential spectrum of neuroinformatics data, and we expect eventually to incorporate many more kinds of information. Here lies a challenge: as the variety of the data managed by a system progresses, the object classes and the relationships (associations) between them increase steadily in number. Thus, when a new class is introduced, there is potentially an association with each existing class. In addition, some classes, such as anatomical structures, may have self-associations (recursive associations). Thus, the thalamus contains, among other structures, the ventrolateral nucleus, while itself being a part of a higher-level structure, the diencephalon.

For the purposes of the rest of this article, we assume that the system is implemented using an object-relational engine, such as Illustra/Informix Universal Server or Oracle version 8. With a straightforward design approach, each class/entity would be represented as a table, while associations between classes would be represented as "bridge" tables with foreign keys referencing one or more entity tables. Often subclasses might need to be derived from a parent class, to store additional, specific information.

With highly heterogeneous data such as is typical of NS data, this approach eventually yields a significant number of classes as well as a complex class hierarchy. More important, the bridge tables can get potentially unmanageable because, with M classes, there are potentially $^{M}C_2$ bridge tables for binary relationships alone (ignoring the possibility of recursive relationships). We must therefore seriously consider ways of simplifying the schema.

The Object Dictionary Approach

A well-known approach, which we term the Object Dictionary (OD) approach, solves the problem partially. (This technique, we believe, was pioneered by Tom Slezak's team in the course of the Lawrence Livermore chromosome 19 mapping project (11). It was subsequently adopted in production systems such as version 5 of the Human Genome Database (12), as well as in DNA Workbench, a package to manage physical mapping data within a chromosomal region (13).) In the OD approach, all classes within the system are children of a parent "Object" class. The **Objects** table contains information on every "object" (class/subclass instance) within a system, with each row typically containing at least the following information: a machine-generated ID, object name and object class ID. (The last is a foreign key into a **Classes** table.) The details of a particular object are found in a class-specific table whose structure is specific to the object's class or subclass, and which is related one-to-one to the Objects table.

One advantage of the OD approach is that, because all object names and definitions are stored in one place, one can create supporting tables (e.g., synonym / keyword tables) to build search tools that have some semblance of intelligence. (Synonyms occur very commonly in NS data: the terms 5-HT, serotonin and 5-hydroxy-tryptamine refer to the same neurotransmitter molecule.) It is unreasonable to insist that most users of an NS database specify the class of object along with the object name in a query, when the name is often unique enough. (For example, "muscarinic" can only refer to a receptor class, and amacrine refers only to a class of retinal neurons.) Only if the term specified by the user is ambiguous is it necessary for the system to display all likely candidates, and force the user to select one.

The OD approach is particularly useful for managing binary associations. Instead of numerous bridge tables for each pair of object classes, we have a single Associations table with at least three columns: Object ID 1, Object ID 2, description of relationship. (In an archival database that gathers information from multiple sources, there is typically a fourth column, a citation/reference.) Only a single

Associations table is needed because the class of any object referenced in the first two columns can always be looked up in the Objects table. For retrieval efficiency in circumstances where a relationship description can have an inverse, the row may be duplicated with the values in the first two columns reversed. To cite a common example in genomic data, a row asserting that object 123 is a sub-clone of object 65 (<123, 65, "sub-clone">) will have a row making the reciprocal assertion (<65, 123, "parent clone">).

The Problem of N-ary Associations

With neuronal data, however, associations between objects are typically not binary but N-ary (where N is greater than 1). For example, consider the following information on the neurons of the nigrostriatal pathway (whose function is impaired in Parkinsonism):

Neurons:	Nigrostriatal
Anatomical Origin:	
Location:	Pars Compacta of Substantia Nigra
Projecting To:	Corpus Striatum
Neurochemical Released:	Dopamine
Receptor:	D2
Electro-physiological Function:	Inhibitory
Neurons that provide Input (Afferents):	Pars Reticulata of Substantia Nigra, Striato-Nigral
Neuron Projected To (Efferents):	Striato-Pallidal neurons, Striato-Striatal neurons

The information in the example above may be regarded as multi-axial, where receptor, anatomical site, neuronal type etc. comprise the axes, and we must now consider how to represent it within a database schema. First, note that it is not advisable to represent the axes as attributes of a "Neuron" class. Such an approach might have been permissible in a database with the primary focus on neurons. However, in a database that stores information on a variety of objects, this approach introduces an asymmetry by implicitly making neurons first-class entities and others second-class. Each axis mentioned above refers to objects which, depending on the perspective of particular users, may be as or more important than neuronal types. Thus, some queries may not be directed at the Neuron class at all: e.g., a user may wish to retrieve a list of anatomical locations where D2 receptors are found. Such users might advocate creating a "Receptors" table, and storing this information with the neuronal type as one of the fields instead.

Therefore, a better way to represent this information is as **associations**. One method of representing multi-axial associations, widely deployed in data warehouse design for business applications, is the "Star" Schema (14). In a star schema, a central "Facts" table, which stores one or more quantitative columns plus several foreign key columns, is related many-to-one to multiple "Dimension" (class) tables. (The phrase "star" refers to the appearance of the schema diagram, with the many-to-one links radiating out from a central facts table.) Each class table stores information on entities in a single axis. This design has proved valuable in situations such as analyzing sales information by territory, salesman, product, product category, volume, and so on. For NS data, however, star schema design is unlikely to be usable without considerable modification, for several reasons:

1. The axes describing neuronal data are **not strictly orthogonal** (i.e. independent). For example, receptors and neurotransmitters are inter-related; given knowledge of the receptor, a domain expert automatically knows the transmitter involved (though the converse is not true).

2. The **number** of axes relevant to a particular fact is variable: some attributes may not be known, and others may be irrelevant for certain instances of data.

3. The **nature** of axes is also unlikely to be static (i.e. unchanging) in a rapidly evolving field. An axis that could be added to the above list, for example, is expression during different phrases of embryonic life.

4. The same object class may appear in more than one axis. In the example, the "Neuron" class appears in three different categories - Neurons, Efferents, and Afferents.

5. Certain axes may have **multiple object instances**. In the example, the nigrostriate neurons receive inputs from multiple neurons and generate outputs.for multiple neurons as well. For the NS, in fact, most of the axes are likely to be multi-valued. (For example, a single neuron has multiple classes of receptors on its surface, and many kinds of neurons are known to release more than one neurochemical simultaneously.) In a normalized relational database design, columns of a table must be atomic and not multi-valued, and so multi-valued data must be factored out into separate tables.

6. Retrieval of data along some axes sometimes involves sophisticated algorithms rather than simple table lookups. The most well known examples in biological data are sequence similarity (determined by algorithms such as BLAST (15) and

110

the information-retrieval metric of Wilbur (16), which measures similarity of two bibliographic citations based on textual content. [1]

7. Within a single axis, entities may be inter-related through recursive relationships of the parent-child type. This complicates the query process because of the need to "explode" a query object instance, retrieving all its children prior to scanning the association data. E.g., in the example above, the pars compacta is part of the substantia nigra, which is part of the mid-brain. To process a query that asked for anatomical locations of various receptors in the mid-brain, one would first have to retrieve all "child" anatomical sites within the mid-brain and then search the association data against this set of child sites.

A general representation of N-ary Associations

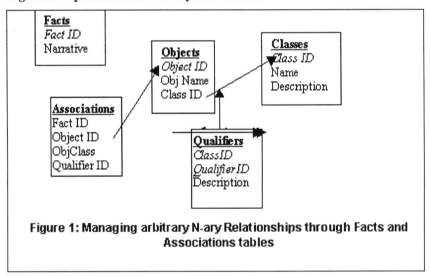

Figure 1: Managing arbitrary N-ary Relationships through Facts and Associations tables

The sub-schema we propose to handle the case of N-ary associations (where N, the number of axes, varies) is shown in fig. 1. (In this figure, table names are Bold/Underlined, while primary keys are in italics. Arrows point from a foreign key to a primary key.) The **Classes** and **Objects** tables have been mentioned earlier in connection with the Object Dictionary approach. The **Facts** table stores a unique identifier, the Fact ID, and a textual narrative of the fact (for reasons described shortly). There is a one-to-many link between the Facts table and a **Citations** table (not shown in the figure).

---

[1] While some systems, such as NCBI's Entrez, store pre-computed sequence and citation similarity scores for efficiency, such pre-computation must be done each time new sequences or citations are added to the database. Such pre-computation is justifiable only if the primarily purpose of the database is to assist similarity searching (as in Entrez).

The **Associations** table describes the object instances linked to the fact, with one row per object instance for the same fact ID. The **Qualifier** field is a descriptor from a controlled-vocabulary table, that describes an aspect of an association. The set of permissible qualifiers is determined by the class of a particular object instance. Examples of qualifiers for neurons are "primary", "afferent" and "efferent". Examples of functional effects are "excitatory", "inhibitory", "autoreceptor negative feedback".

The example data would be represented in the Associations table as illustrated in Table 1. (We assume a fact ID of 100 and, for simplicity, use the Object / Class / Qualifier descriptions rather than IDs.)

Fact ID	Object	Class	Qualifier
100	Nigrostriate	Neuron	Primary
100	Pars Compacta, S.Nigra	Location	Soma
100	Corpus Striatum	Location	Eff.Axon
100	Dopamine	Neurochemical	Transmitter
100	Receptor	D2	Efferent
100	Inhibition	Electr.Function	
100	Striato-pallidal	Neuron	Efferent
100	Striato-striatal	Neuron	Efferent
100	Striato-nigral	Neuron	Afferent
100	Pars Reticulata, S.Nigra	Neuron	Afferent

Table 1: Capturing Information on Nigrostriatal Neurons

The Qualifier field can readily capture the semantics of an association in the case of binary relationships. Thus, to record the assertion that the neurochemical "substance P" is co-released as a modulator with the chemical serotonin (a neurotransmitter), we would need two rows. One row would have the Object ID/Qualifier values "serotonin" and "transmitter", the other would have "Substance P" and "modulator". Reciprocal binary relationships are also readily captured without the need to store the same fact reciprocally. Thus, to record the assertion, "structure X is contained within structure Y", one uses two rows with the Object/Qualifier pairs <"X",."contained"> and <"Y","contains">.

However, for the arbitrary N-ary fact, the objects are linked to each other conceptually like the nodes of a semantic network. The Qualifiers act somewhat like the edges in the network, but they are not fully satisfactory for this purpose given the current structure. Therefore, it is often hard to reconstruct the semantics of an N-ary fact given the data in the Associations table alone. (Even in cases where it is not hard, it generally requires several computational steps.) To avoid this (generally unneeded) computation, the "Narrative" field in the Facts table stores an explicit textual description of the N-ary fact for the user's perusal.

By an analogy with Information-Retrieval methods, the Associations table may be regarded as an **index** (17) to the narrative text, for the purpose of rapid retrieval. The only difference between the Association table and the inverted files created by free-text indexing engines (e.g., for Web-searchable document collections) is that the index-term vocabulary is more controlled with the Associations table. The similarity, however, is that, in both cases, complex Boolean retrieval (e.g., list all neurons where Dopamine has an inhibitory role) requires set operations, such as Union. Intersection, and Difference, on subsets (projections/ selections) of the Associations table with each other. Relational set operations are computationally less efficient than the equivalent AND, OR and NOT operations that would have been needed with, say, a classical Binary-Relationship table, but the plus feature is flexibility and a simple structure. (For example, multiple object instances on a single axis do not need to be managed through separate many-to-one related tables.) Also, in practice a significant proportion of queries tend to be based on a single axis rather than multiple ones. Such queries can be answered by locating the fact IDs corresponding to a particular class instance, and then simply returning the narrative for those IDs.

Managing Hierarchical Associations

While the Associations table can manage arbitrary N-ary data, that does not mean that every association in the database must be stored this way. Use of the Associations table should be restricted to represent highly **heterogeneous** facts (where both the number and the nature of the axes vary greatly). Facts best managed in the orthodox fashion include **parent-child** relationships, a special category of binary relationship. These are seen quite commonly in the NS, for example, with receptors (which have subtypes), and anatomical structures (which have sub-structures).

We have mentioned earlier the need to preprocess queries accessing hierarchical data. This way, a query specified at a coarser level of granularity must also be able to retrieve facts stored in the database at a finer level of granularity, without having to store facts redundantly at multiple granularity levels. Standard transitive-closure algorithms for this purpose have been well-researched for the "Bill of Materials Problem" (18). Limited transitive-closure support will be provided in SQL-3 (19).

Managing the Object Class Hierarchy

The NS has many classes of data, and some of these classes form a natural hierarchy. Consider, for example, molecules with a role in the NS. The smallest such molecules are the neurotransmitters, odor molecules and "second messengers" such as cyclic Adenosine Monophosphate. The neuromodulators such as the enkephalins (composed of a small number of amino-acids) and neurohormones (e.g., endorphins) are somewhat larger. At the macromolecular end, we have various protein molecules such as enzymes, ion channels and receptor molecules, along with their corresponding nucleotide sequence/s. All of these are sub-classes of the class "molecule", and the nature of information to be recorded against each subclass varies. Thus, simple structural formulas (as vector graphics) are useful for smaller molecules, while the larger protein molecules need to have secondary or tertiary structures recorded when available. For enzymes, one would like to know about the reactions they catalyze.

Similarly, for anatomical structures, one can descend from gross anatomical structures (e.g., the cerebellum), to sub-structures (such as nuclei), to individual layers of cells and the neuronal cell types within a layer.

However, there are some practical problems in directly porting the object-oriented inheritance paradigm to design of a schema for NS objects.

- While the number of classes is large, many of them will have only a modest number of actual object instances. (For example, even though different information needs to be recorded on a neurotransmitter versus a neuromodulator, the currently known neurotransmitters plus neuromodulators number less than a hundred.) It may seem like overkill to create separate tables as part of a complex hierarchy for such few instances, because the added design effort is not sufficiently amortized in terms of the volume of the data that is better managed.

- In the context of NS data, we are using the phrase "class" more loosely than, say, Java programmers do. In programming parlance, a class refers to a particular **data structure** as well as the permissible **operations** (member functions, methods) that can be defined for that class. However, for most NS "classes", one cannot really think of any applicable operations other than simple input and output.

Simplifying Class Management through EAV Design

When the number of instances of a class is expected to be numerous enough in a production schema, the standard Object Dictionary approach (with class-specific tables holding object descriptions, as mentioned earlier) is adequate. When it is not, we must consider alternative design approaches.

One approach that is popular for modeling highly heterogeneous data is the Entity-Attribute-Value (EAV) approach. (Some authors substitute "Object" for Entity.) Attribute-Value pairs as a means of describing an object were pioneered in

AI in the form of LISP association lists (20). Subsequently, they have been used for applications as diverse as Clinical Data Repositories (21) and are the basis for the structure of Web cookies (22) and the Microsoft Windows Registry, as well as various tagged formats for data interchange, such as ASN.1 (23). We have implemented an EAV design in ACT/DB a production system for managing clinical trials data (24).

Conceptually, in an EAV design, we have a table with three columns, Entity ID (Object ID), Attribute (or Attribute ID) and the Value for the attribute. Thus, one describes an object through several rows in a table, one row per Attribute-Value pair. Thus, to record information about the enzyme "Cholinesterase" (object ID 250), which destroys the neurotransmitter Acetylcholine by hydrolysis, one might have the following rows:

<250, "Substrate", "Acetylcholine">
<250, "Action", "Hydrolysis">
<250, "Inhibited-by", "Organophosphorus compounds">
...

Attributes are analogous to fields in a conventional table. In the examples, all values are strings. In practice, relational databases are strongly typed. Rather than have a single EAV table (with the Values coerced into string form) it is often advantage to segregating values into separate EAV tables based on their intrinsic datatype, for the following reasons.

• One can improve retrieval performance by creating an index on a combination of the attribute and value fields, to permit queries based on the values of the attribute. (Indexes on numeric fields stored in string form are essentially useless because of a different collating order.)

• One can create EAV tables to hold values that are long text (e.g., sequences, or PDF data) or BLOB data (such as vector graphics or photographic images). The ACT/DB system uses six EAV tables for integer, float, string, date, long text and BLOB data.

The success of an EAV system depends critically on the supporting metadata. We describe our metaschema in Figure 2.

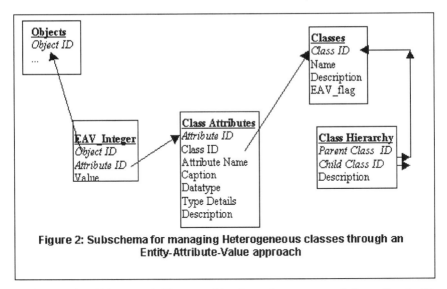

Figure 2: Subschema for managing Heterogeneous classes through an Entity-Attribute-Value approach

In fig. 2, the Objects and Classes tables have been repeated from fig. 1. The Classes table has an extra field, EAV_Flag, a Boolean that records whether the implementation of this class is EAV (if true) or conventional (if false). EAV makes sense only for heterogeneous data, and there is no reason why a homogenous class with a large number of object instances should not be implemented conventionally.

For each class, a description of the applicable attributes is stored in the **Class_Attributes** table. Each Attribute has a name, a caption (for display to end-users as a label in a form) a data type, a serial number (that determines order of presentation) and a description (documentation). For BLOB datatypes, the "Type Details" field records the kind of data (e.g., mime-type) that an attribute represents, to assist management by application software. The Class Attributes table not only documents the class definition, but is **actively** consulted by the system for construction of input screens and generation of formatted output..

The actual data for an object instance is stored in the EAV tables. (Only one of these, a table for integers, is shown in the figure.) The table **Class_Hierarchy** records the parent-child relationships between Class definitions (not class instances: these are stored in an Object_Hierarchy table as described earlier). This table is consulted when the user specifies a query based on a super-class that might encompass sub-classes as well.

Limitations of the EAV model

The simplicity of EAV comes with a price, namely, a performance penalty. The physical representation of a class is quite different from its logical view as seen by the user. Assembling all the "columns" associated with a particular class object involves consulting the Class Attributes table and then gathering data from the appropriate EAV tables. More important, complex Boolean query of classes requires set operations (as in the case of the general N-ary relationship structure described earlier).

In brief, EAV is not panacea. In any production system, data will be stored in a mixture of conventional or EAV forms. Also, the schema may mutate, with a particular physical classes being switched, as generality and efficiency are traded off. A production EAV system should endeavor to hide the EAV or non-EAV nature of the class data from the user, so that it always appears conventionally structured. This introduces complexity when querying the Class data. The user/programmer must be shielded from having to remember how a class is physically represented, and a query generator must be used to generate the appropriate SQL based on which class/es of objects are specified in a complex query. (We have created a query generator for patient data in a clinical trials database that works on a combination of conventional and EAV data; this work is described in (25).).

Conclusions

Our proposed database schema for managing heterogeneous data is a significant departure from conventional approaches. It is suitable only when the following conditions hold:

- The number of classes of entity is numerous, while the number of actual instances in most classes is expected to be very modest.

- The number (and nature) of the axes describing an arbitrary fact (as an N-ary association) varies greatly.

We believe that nervous system data is an appropriate problem domain to test such an approach.

References

1. Shepherd G, Mirsky JS, Healy MD, et al. The Human Brain Project: Neuroinformatics tools for integrating, searching and modeling multidisciplinary neuroscience data. Trends in Neurosciences (in press).
2. Shepherd GM, Healy MD, Singer MS, et al. Senselab: a project in multidisciplinary, multilevel sensory integration. In: Huerta. SHKMF, ed. Neuroinformatics: An Overview of the Human Brain Project,. Mahwah, NJ: Lawrence Erlbaum Associates, Inc., 1997: 21-56.
3. Koslow S, Huerta Me. Neuroinformatics: An Overview of the Human Brain Projects. Mahwah, NJ: Lawrence Erlbaum Associates, 1997.
4. Healy MD, Smith JE, Singer MS, et al. Olfactory Receptor Database (ORDB): A resource for sharing and analyzing published and unpublished data. Chemical Senses 1997;22:321-326.
5. Peterson B, Healy M, Nadkarni P, Miller P, GM. S. ModelDB: An environment for running and storing computational models and their results applied to neuroscience. J. Amer.Informatics Assoc 1996;3(6):389-398.
6. Bower JM, Beeman D. The Book of Genesis. New York: Springer-Verlag, 1995.

7. Mirsky JS, Nadkarni PM, Hines M, Healy MD, Miller PL, Shepherd GM. A framework for informatics support of computer-based neuronal modeling: imposing order in a complex domain. (in preparation).

8. Hines M. The NEURON simulation program. In: Skrzypek J, ed.   Neural Network Simulation Environments. Norwell, MA: Kluwer Academic Publishers, 1993:

9. Mirsky JS, Nadkarni PM, Healy MD, Miller PL, Shepherd GM. Database tools for integrating neuronal data to facilitate construction of neuronal models. Journal of Neuroscience Methods (in press).

10. Skoufos E, Nadkarni P, P M. Using an Entity-Attribute-Value Data Model to Store Evolving Experimental Data from Biomedical Research. (in preparation).

11. Slezak T, Wagner M, Yeh M, et al. A Database System for Constructing, Integrating, and Displaying Physical Maps of Chromosome 19. In: Hunter L, Shriver BD, eds. Proceedings of the 28th Hawaii International Conference on System Sciences. Wialea, Hawaii: IEEE Computer Society Press, Los Alamitos, CA, 1995:14-23.

12. Fasman KH, Letovsky SI, Cottingham RW, Kingsbury DT. Improvements to the GDB Human Genome Data Base. Nucl. Acids. Res. 1996;24(1):57-63.

13. Nadkarni PM, Cheung K-H, Castiglione C, Miller PL, Kidd KK. DNA Workbench: A Database Package to Manage Regional Mapping. Journal of Computational Biology 1996;3(2):319-329.

14. Kimball R. The Data Warehousing Toolkit. New York, NY: John Wiley, 1997.

15. Altschul SF, Gish W, Miller W, Myers EW, Lipman DJ. Basic Local Alignment Search Tool. J. Mol. Biol. 1990;215:403-410.

16. Wilbur WJ, Yang Y. An analysis of statistical term strength and its use in the indexing and retrieval of molecular biology texts. Computers in Biology & Medicine 1996;26(3):209-222.

17. Salton G. Automatic Text Processing: the transfomation, analysis, and retrieval of information by computer. Reading, MA: Addison-Wesley, 1989.

18. Goodman N. Bill of Materials in Relational Database. InfoDB 1990;5(1):2-13.

19. Melton J. (editor). ISO/IEC SQL Revision. ISO-ANSI Working Draft Database Language SQL (SQL3). New York, NY: American National Standards Institute, 1992:

20. Winston PH. Artificial Intelligence. (2nd ed.) Reading, MA: Addison-Wesley, 1984.

21. Huff SM, Haug DJ, Stevens LE, Dupont CC, Pryor TA. HELP the next generation: a new client-server architecture. Proc. 18th Symposium on Computer Applications in Medical Care. Washington, D. C.: IEEE Computer Press, Los Alamitos, CA, 1994:271-275.

22. Dwight J & Erwin M (eds). Special Edition: Using CGI. Indianapolis, IN: Que Corporation, 1996.

23. Huff SM, Rocha RA, Solbrig HR, W BM, P SS, M S. Linking a Medical Vocabulary to a Clinical Data Model using Abstract Syntax Notation 1. (in press).

24. Nadkarni PM, Brandt C, Frawley S, et al. Managing attribute-value clinical trials data using the ACT/DB client-server database system. Journal of the American Medical Informatics Association 1998;5(2):139-151.
25. Nadkarni P, Brandt C. Data Extraction and Ad Hoc Query of an Entity-Attribute-Value Database. Journal of the American Medical Informatics Association in press.

10 THE MOUSE GENOME DATABASE AND THE GENE EXPRESSION DATABASE: GENOTYPE TO PHENOTYPE

Janan T. Eppig, Joel E. Richardson, Judith A. Blake, Muriel T. Davisson, James A. Kadin, and Martin Ringwald

The Jackson Laboratory, Bar Harbor, ME 04609

Introduction

Two important resources for mouse biological and genomic data and analysis exist, and continue to expand at The Jackson Laboratory (U.S.A.). These resources, the Mouse Genome Database (MGD) and the Gene Expression Database (GXD), have as their long term goal the facilitation of research through access to integrated genomic structural data, gene expression information, and phenotypic descriptions. The World Wide Web (WWW) provides a method for easily navigating these integrated data resources to address complex questions of biological importance (Figure 1).

The goals for the MGD and GXD resources are 1) to provide high quality, elemental genomic data suitable for analysis and integration with other biological data, 2) to develop software for data importation, analysis and display, and 3) to develop easy-to-use flexible interfaces for access by the scientific community. Emphasis is on acquisition of primary data, since these are most suitable for analysis and combination with new data.

120

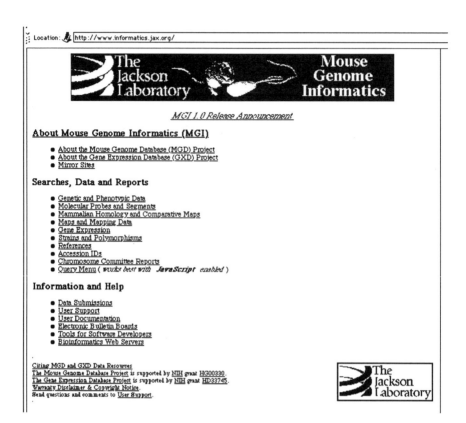

Figure 1. Mouse Genome Informatics Home Page. The Mouse Genome Informatics World Wide Web site (http://www.informatics.jax.org) serves as an integrating point of access for the Mouse Genome Database (MGD) and the Gene Expression Database (GXD). This unifying WWW site provides users with seamless easy access to information on the genetics and biology of the laboratory mouse.

Mouse Genome Database (MGD)

The Beginnings

The Mouse Genome Database developed as an outgrowth and extension of various information resources, both paper and electronic, on the laboratory mouse. The tradition of pre-publication and anecdotal information exchange in the mouse genetics community fostered the early development of data compilations and consensus map building.

Early compilations of mouse genetic data pre-date electronic databases. The first gene description catalog for the mouse was published from The Jackson Laboratory in 1941 by Dr. George Snell [1]. Dr. Margaret Green should be credited as the developer of the first mouse genetics database when, in the 1950s, she began an index card file system delineating published and personally communicated results of experimental crosses. These formed the basis of early versions of a composite mouse genetic linkage map [2-6]. Later, she also compiled descriptions of mutant and polymorphic genes in the mouse, the centerpiece for the first edition of *Genetic Variants and Strains of the Laboratory Mouse* [7] published in 1981. These gene description data (later known as the Mouse Locus Catalog, MLC) were maintained as a word-processing document.

In the 1980s, GBASE (Genomic Database of the Mouse), the first online resource of mouse genomic information, was developed by Drs. Roderick and Davisson [8]. GBASE provided a single menu for user access to three independent data sets: Locusbase, an Ingres database with a character cell interface, contained summarized mapping data initially populated from Dr. Green's cards; MATRIX, with a command-line interface, contained strain-by-locus allele data; and MLC, with an IRX text searching interface, contained synoptic descriptions of genes.

In 1989, the first incarnation of the *Encyclopedia of the Mouse Genome* (*Encyclopedia*), a suite of software tools for viewing mouse genetic data, was developed through the collaborative work of Drs. J.H. Nadeau, L.E. Mobraaten, and J.T. Eppig [9]. This software provided intuitive, graphical user interfaces for browsing genetic and cytogenetic maps, associated references, notes, and gene descriptions. Its purpose was to provide simultaneous access to information derived from different database sources and to provide means of querying those various sources using a single computer mouse 'click'. The *Encyclopedia* offered graphical browsing of the Chromosome Committee reports for the mouse and the MIT Genome Center SSLP maps under both UNIX and Macintosh operating systems.

Also during the 1980s, a number of domain specific databases were developed to fill specific research needs and produce periodic publications. Among these were a database containing primary haplotype mapping data to support linkage analysis and map drawing programs, a database of probes, clones, and molecular markers characterizing these new molecular reagents and associated RFLP data [c.f., 10], and a database of homology relationships between the mouse and other mammalian species [11]. A compilation of the characteristics of 728 laboratory mouse strains also was initiated during this period by synthesizing information from *Cancer Research* listings about inbred strains dating back to 1952 [12].

In 1992, the Mouse Genome Database was initiated with NIH funding. Several significant challenges came with MGD's early development. Initial requirement analysis, database design, and implementation had to be accomplished in a backdrop of continued maintenance of the existing GBASE online resource. The pre-existing databases had to be assimilated and, most importantly, integrated. Although this legacy provided a foundation of data already in electronic form, the data existed on different platforms, in different data management systems, and with fundamentally different data paradigms and organizations. Further, these databases contained varying amounts of overlapping information (*e.g.*, gene symbols, references), requiring extensive programmatic and manual comparisons, and data reconciliation, and data resolution.

The Present

In 1994, the first public release of MGD appeared on the WWW, although full integration of all pre-existing systems was not yet complete [13-16]. Since then, MGD has continued to evolve its underlying structure to maximize data integration, as well as expand its data coverage. Data volume has grown continuously; many new types of data sets have been added, and software developments have enhanced data access and provided new tools for users to explore data relationships. MGD is updated continuously with new data added daily. Software changes to the database schema, enhancements to the user interface, and new analysis and display tools are developed, tested and made available through our WWW site at 3-4 month intervals.

MGD includes structural genomic and phenotypic data about the mouse. Currently, the data sets include:

- Genetic marker characteristics: gene identification and definition data; gene symbol, name, alleles, nomenclature
- Mapping data: experimentally generated genotypic data from linkage crosses, recombinant inbred strain experiments, and many other genetic mapping techniques
- Physical mapping data: probe vs. YAC hit/miss data; STS content data
- Comparative mapping data: for mouse and >50 other mammalian species
- Maps: genetic, cytogenetic, physical maps of mouse; comparative maps between mouse and other mamals
- Molecular segments: clone, probes, primers, ESTs; their characteristics and sequence links
- Polymorphisms: visible, biochemical, RFLP or PCR variant polymorphisms among strains
- Phenotypic data : descriptions of genes, mutations, and their function
- Inbred strains: descriptions of inbred strains of mice, including their unique quantitative characteristics.
- Chromosome Committee reports: for viewing or downloading (1995 to present).

Data acquisition for MGD includes curation of the published literature, data downloads from genome centers, and individual researcher data submissions and annotations. Current data downloads are from a number of DNA mapping panel providers, including the EUCIB (European Collaborative Interspecific Backcross) panel, The Jackson Laboratory panel, and the NCI-Frederick Cancer Research and Development Center panel; the Whitehead Institute MIT (mouse physical maps and data); the I.M.A.G.E. (Integrated Molecular Analysis of Genomes and their Expression) consortium (clones and libraries); and the Washington University/Howard Hughes Medical Institute mouse EST project (EST data).

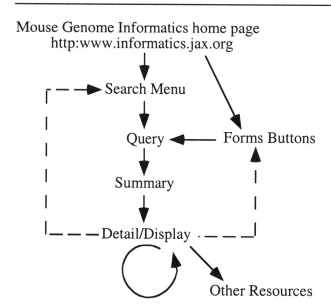

Figure 2. General WWW navigation of MGD. From the homepage, a topic area for initial search is chosen, such as 'Genes, Markers, and Phenotypes'. A search menu for this topic area offers users several query forms and reports available. A query is completed, a summary of the data sets fulfilling the criteria are returned, and the user can then display the details for those data of interest. Hypertext links from the detail pages link to additional information within MGD relevant to the data being viewed and to external databases containing related data. Users can redirect their queries by jumping to the search menu. A menu of query form buttons also is available as a shortcut for more advanced users.

The WWW interface to MGD provides a variety of search options, with search results presented as summaries, tabular details, or graphical displays. The users' general navigation paradigm involves selecting a starting point for data query, focusing the search by optionally filling in fields in a query form, and further selecting specific data sets from those that satisfy the query. Detail data pages contain hypertext links to other data in MGD or in external data resources, where relevant, further enriching the information available to users (Figure 2). In addition, a number of pre-generated reports are available for frequently requested large data lists, including complete lists of genetic markers and complete tables of mouse-human and mouse-rat gene homologies.

Maps are viewed in three formats: in the browser window (Web map) with hypertext links to gene data; using the interactive *Encyclopedia of the Mouse Genome* software tool; or as a publication quality map printed from a PostScript file. Each genetic and comparative map is generated 'on-the-fly' based on user defined parameters specifying data set, marker types or classifications, region to be displayed, and whether to show homologous genes for another species. Cytogenetic and physical maps currently are only available as Web maps.

MGD strives to provide the scientific community with the most up-to-date information in an easy-to-use environment. Special requests for information not easily retrieved through the WWW interface can be accessed directly using the public SQL server. Alternatively, our User Support team will do specialized user queries on request.

Gene Expression Database (GXD)

Differential gene expression generates complex spatio-temporal networks of gene and protein interactions. The laboratory mouse, as an important animal model in the study of human disease, is being used extensively in gene expression studies. Emerging high throughput methods make it possible to analyze thousands of genes simultaneously for expression in different tissues. Such experiments will provide global expression profiles that can be used to guide focused expression studies using more conventional expression assays, such as Northern and Western blot, RNA *in situ* hybridization, and immunohistochemistry, to determine what transcripts and proteins are produced by specific genes, and where and when these products are expressed at the cellular level. The goal of the Gene Expression Database [17, 18] is to support the storage and analysis of all these data, with the initial concentration being mouse embryonic development.

In February 1996, the GXD Index, a view into the literature on mouse embryonic gene expression, was made available through the WWW. This initial offering provides users with a searchable index of research reports documenting data on endogeneous gene expression during mouse development. For each scientific publication, the Index includes the genes studied, the embryonic ages analyzed, and the expression assays used. The GXD Index is integrated with MGD to foster a close link between genotype, expression, and phenotype information.

In February 1998, a 'cDNA and EST Expression' search was added. Unlike the MGD 'Molecular Probes and Segments' search, the GXD format enables better searching on source data associated with expressed sequences. Particularly, tissue, age, cell line, and library are prominent search fields.

In June 1998, the first large-scale expression data set was searchable through the 'Gene Expression Data' query form. These data, generated by T. Freeman (Sanger Centre, U.K.) include RT-PCR assays for 517 genes in 45 mouse tissues from 6-8 week old animals and from 15 day old embryos. Assay data include primary data on each sample prepared, expression profiles for each sample in each experimental gel, and images of the gels.

Importantly, although only limited expression data are available at this writing, GXD is poised to undergo rapid data expansion. The database is now implemented to capture and make available other datasets and several other types of expression information. Expression patterns are described by a comprehensive dictionary of anatomical terms that has been developed in collaboration with Drs. Bard and Kaufmann of the University of Edinburgh and Drs. Davidson and Baldock of the MRC, Western General Hosptial, Edinburgh. Further, editorial interfaces for capturing and updating assay data from publications and image scanning are in place.

Some expression data will be acquired from annotation of the literature by database editors. However, because only a fraction of the gene expression data that a laboratory generates actually appears in published form, it is anticipated that GXD data largely will come from electronic data submissions. The Gene Expression Annotator (GEA) has been developed for this purpose and is currently being tested by several laboratories. The Annotator prototype provides important features for capturing standardized descriptions of gene expression data, validating data, and submitting data files. GEA supports a drag-and-drop facility for importation and indexing of image files, a hierarchical look-up list for embryonic anatomy, and links to resources such as MGD and GenBank for verifying nomenclature and describing probes [18].

With the sum of these developments, GXD will quickly expand and new data sets and features can be expected to appear regularly on the Mouse Genome Informatics WWW site.

The Mouse Genome Informatics (MGI) WWW site

MGD and GXD together provide a unique resource for analyzing how the structural genome, through developmental pathways, produces observed phenotypes (Figure 3). These databases are tightly integrated to enable comprehensive analysis of genotype, expression, and phenotype data. Further, as a practical consideration, because MGD already contains many data types that need to be shared with GXD (*e.g.*, data on genes, molecular probes, inbred and mutant strains, references), efficiency is gained by coordinating maintenance of these data.

126

Genotype　　　　**Expression**　　　　**Phenotype**

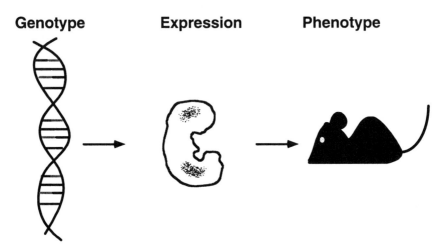

Figure 3. The Mouse Genome Database (MGD) provides genetic, genomic, and phenotypic data on the mouse. These represent the two ends of the spectrum illustrated here. The Gene Expression Database (GXD) bridges the molecular and developmental aspects of translating genotype to phenotype. MGD and GXD are accessible through a common WWW site at http://www.informatics.jax.org.

MGD and GXD provide user access through a common interface, the Mouse Genome Informatics WWW site (http://www.informatics.jax.org). This site was developed with the premise that biological scientists interested in accessing data on the genetics and biology of the laboratory mouse prefer to access data in a coordinated, integrated system. The MGI WWW site provides seamless access to MGD and GXD, with extensive hypertext linking between the data sets, such that the user can traverse all data without intentionally having to switch from a MGD to a GXD view, and vice versa. Importantly, the MGI WWW site will allow future expansion into other areas of mouse biological data, such as mouse tumor information, which is slated to be released at our site later this year.

Future

MGD and GXD will continue to grow, expand, and evolve. Many new challenges are anticipated as both high throughput and exquisitely sensitive methodologies continue to develop, allowing new and exciting biological data to be collected. As well, evolving computer technologies will give us the ability to analyze and display these data in flexible graphical ways. The mouse has become a pivotal animal model system. New emphasis is being placed on sequencing its genome and on generating new mouse mutants by homologous recombination and saturation mutagenesis. Comprehensive genetic, expression, and phenotypic analysis of these mutants will provide unprecedented insights into the mechanisms that underlie normal development and disease. MGD and GXD will continue to play an important role in the acquisition and analysis of these critical biological data for the laboratory mouse.

Acknowledgements

MGD is funded by NIH grant HG00330; GXD is funded by NIH grant HD33745. We thank the entire database staff for thier contributions to the success of these projects: R.M. Baldarelli, J.S. Beal, D.A. Begley, R.E. Blackburn, J.J. Bobish, D. W. Bradt, C.J. Bult, N.E. Butler, G.T. Colby, L.E. Corbani, G.L. Davis, C.J. Donnelly, D.P. Doolittle, K.S. Frazer, J.C. Gilbert, L.H. Glass, P.L. Grant, D.M. Krupke, M. Lennon-Pierce, L.J. Maltais, M.E. Mangan, M.E. May, M.G. McIntire, J.J. Merriam, J.E. Ormsby, R.P. Palazola, S. Ramachandran, D.J. Reed, S.F. Rockwood, D.R. Shaw, L.E. Trepanier, P. G. Trepanier, H. Zhou.

References

1. Snell, G.D. Genes and chromosome mutation. In: *Biology of the Laboratory Mouse*, 1st Edition, Snell GD, ed. McGraw-Hill, New York. pp 234-247, 1941.
2. Green, M.C. The linkage map of the mouse. Mouse News Lett 47:16, 1972.
3. Green, M.C. The linkage map of the mouse. Mouse News Lett 49:17, 1973.
4. Green, M.C. The linkage map of the mouse. Mouse News Lett 51:7, 1974.
5. Green, M.C. The linkage map of the mouse. Mouse News Lett 53:9-10, 1975.
6. Green, M.C. The linkage map of the mouse. Mouse News Lett 55:6, 1976.
7. Green, M.C. Catalog of Mutant Genes and Polymorphic Loci. In: *Genetic Variants and Strains of the Laboratory Mouse*, 1st Edition, Green MC, ed. Fischer Verlag, Stuttgart. pp 8-278, 1981.
8. Doolittle, D.P., Hillyard, A.L., Davisson, M.T., Roderick, T.H. and Guidi, J.N. GBASE - The genomic database of the mouse, Fifth International Workshop on Mouse Genome Mapping, Lunteren, Netherlands. p 27, 1991.
9. Eppig, J.T., Blackburn, R.E., Bradt, D.W., Corbani, L.E., Davisson, M.T., Doolittle, D.P., Drake, T.E., Frazer, K.S., Grant, P.L., Guidi, J.N., Lennon-Pierce, M., Maltais, L.J., Mankala, S., May, M.E., McIntire, M.G., Mobraaten, L.E., Nadeau, J.H., Ormsby, J.E., Reed, D.J., Richardson, J.E., Rockwood, S.F., Roderick, T.H., Sharpe, S.B., Shroder, S.C., Smith, A.G. and Stanley, MlL. The Encyclopedia of the Mouse Genome, an update. Third International Conference on Bioinformatics and Genome Research, Tallahassee. p 73, 1994.
10. Eppig, J.T. Mouse DNA clones, probes and molecular markers. Mouse Genome 91:594-755, 1993.
11. Nadeau, J.H., Davisson, M.T., Doolittle, D.P., Grant, P.L., Hillyard, A.L., Kosowsky, M. and Roderick, T.H. Comparative map for mice and humans. Mamm Genome 1:S461-S515, 1991.
12. Staats, J. Standardized nomenclature of inbred strains of mice: Eighth listing. Cancer Res 45:945-977, 1985.
13. Richardson, J.E., Eppig, J.T., Nadeau, J.H. Building an Integrated Mouse Genome Database. IEEE Engineering in Medicine and Biology 14: 718-724, 1995.
14. Blake, J.A., Eppig, J.T., Richardson, J.E., Davisson, M.T. and the Mouse Genome Informatics Group. The Mouse Genome Database - (MGD) - A

Community Resource. Status and Enhancements. Nucleic Acids Res 26(1): 130-137, 1998.

15. Blake, J.A., Richardson, J.E., Davisson, M.T., Eppig, J.T. and the Mouse Genome Informatics Group. The Mouse Genome Database (MGD). A comprehensive public resource of genetic, phenotypic and genomic data. Nucleic Acids Res 25(1):85-91, 1997.

16. Eppig, J.T., Blake, J.A., Davisson, M.T. and Richardson, J.E. Informatics for Mouse Genetics and Genome Mapping. *Methods: A Companion to Methods in Enzymology* 14:179-190, 1998.

17. Ringwald, M., Baldock, R., Bard, J., Eppig, J.T., Kaufmann, M., Nadeau, J.H., Richardson, J.E. and Davidson D. A database for mouse development. Science 265: 2033-2034, 1994.

18. Ringwald, M., Davis, G.L., Smith, A.G., Trepanier, L.E., Begley, D.A., Richardson, J.E. and Eppig, J.T. The mouse gene expression database GXD. Sem Cell & Devel Biol 8:489-497, 1997.

11 THE EDINBURGH MOUSE ATLAS: BASIC STRUCTURE AND INFORMATICS

Richard A Baldock, Christophe Dubreuil, Bill Hill and Duncan Davidson

MRC Human Genetics Unit, Western General Hospital, Crewe Road, Edinburgh, EH4 2XU, UK.

Introduction

The rapid accumulation of information from *in situ* experiments on the developing mouse embryo, for example hybridization or immunohistochemistry, together with its spatial and temporal complexity demands a spatio-temporal database to allow collation, comparison, analysis and query. The patterns of gene-activity and consequent effects on the tissue development are alternative views of the structure of the embryo which has traditionally been described using morphologically defined structures or *anatomy*. To record this data, in order to understand the complex interactions between genes and the consequent morphogenesis and differentiation, we require a means of mapping the spatial information onto a neutral representation. The most appropriate representation for this task is the mouse embryo itself therefore we are generating 3-D image or *voxel* models of the mouse embryo, initially at each developmental stage defined by Theiler [1], but with the possibility of extension to finer time-steps especially at the earlier stages. The voxel models are 3-D arrays of image values corresponding to a conventional histological section as viewed under the microscope and can be digitally re-sectioned to provide new views to match any arbitrary section of an experimental embryo. The design of the *gene-expression* database [2,3,4,5,6], and the reconstruction methods [7,8], have been presented before and will not be repeated here. In this paper we discuss the underlying design of the spatial aspects of the atlas and database, and some of the associated bioinformatics issues.

130

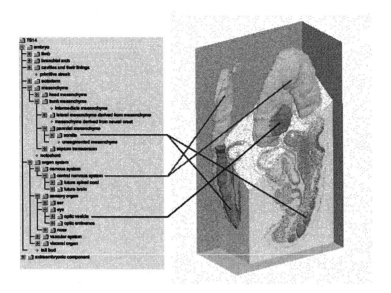

Figure 1: Basic structure of the Edinburgh Mouse Atlas. The standardized and structured anatomy nomenclature is linked (in the database) to 3D images representing the corresponding region in the grey-level 3D reconstruction. This is shown sectioned to reveal the histology.

The purpose of this database is to provide a means of comparing and analyzing gene-expression, or other spatially organized data, within the developing mouse embryo. Anatomy is our "external" description of the embryo while images of expression domains of different genes brought together into a single space-time frame provide an alternative "internal" description and a primary aim of developmental genetics is to understand the "external" description in terms of the internal gene activity. For this reason our database must include an unambiguous mapping between space and anatomy. This implies a standardized anatomical nomenclature as well as a mapping from anatomical term to spatial region or *domain* in the reconstructions. This nomenclature database has been defined and is Internet-accessible (genex.hgu.mrc.ac.uk). Figure 1 shows a graphical view of the structure of the atlas.

The combination of the 3-D voxel models, the standardised nomenclature (and associated lineage and tissue information) and the mapped anatomical domains is known as the *Edinburgh Mouse Atlas* (EMA) and is being developed as a collaboration between the Department of Anatomy, University of Edinburgh and the MRC Human Genetics Unit.

The nomenclature is also the key element for a purely textual description of gene-expression patterns and provides the link between the spatio-temporal gene-expression database under development at the MRC, UK and the gene-expression database (GXD) at the Jackson Laboratory, USA [3,6].

Voxel Models and Database Structure

Database systems for spatially mapped information have been developed and in long use as Geographic Information Systems and a number are now emerging for biomedical research and teaching [9,10,11,12]. The special feature of biomedical atlases is that there is no single correct object, i.e. the underlying structures show natural variation and the processing steps to acquire the data introduces additional systematic and random spatial variation. This raises additional problems of data mapping in order to enable spatial comparison. For gene-expression patterns the basic data arises from images of tissue sections or "whole-mounts" which provide a projection of the whole 3D pattern onto a 2D image. To map this onto a standard embryo it must be possible to find the corresponding section or projection, and then transform or warp the experimental data onto that atlas section or projection. In the future, mapping methods from 3D data, e.g. from confocal microscopy, MRI or section reconstructions, such as those developed for brain-mapping [13] will have to be provided. The simplest reference model which can allow resectioning and mapping of spatial information is a 3D array of image values or *voxels*.

For the purposes of developmental genetics the voxel model should be *interpretation-free*, i.e. it must be possible to map information independently of any other interpretation or description of the embryo (e.g. anatomy). This is important because in general the visible morphological features (anatomy) at any stage will not provide adequate descriptors for the gene-activity which may be related to the development of form not yet visible. For the purposes of the database we have selected the most commonly used histological staining (H&E), and digitized images of the serial sections in order to reconstruct the 3-D image. For the EMA the spatial resolution is determined by the original microtome section thickness which is 2 microns for Theiler stages 1-12 and 6 to 8 microns for older embryos.

Experimental data is mapped onto the models as a list of voxel locations (3D binary image or *domain*). The gene-expression database will thus have a similar structure to that shown in figure 1 for the anatomy. The reference reconstructions act to define the coordinate system onto which the database entries with spatial domains can be mapped. These initially will be the anatomical components and gene-expression patterns but can be extended to any spatially organized data.

The image processing software adopted for this project has been developed by the MRC Human Genetics Unit and is known as *Woolz* [14]. This is implemented in C and is in the public domain (genex.hgu.mrc.ac.uk). The gene expression database has an object-oriented design [15] and we have selected ObjectStore (www.odi.com) as the database management system because it allows the data object to be defined in C++, which provides transparent access to the Woolz objects and procedures for the purposes of update and query.

Atlas Spatial Coordinates

In defining an atlas it is important that the coordinate system make biological sense, provide help for navigation and allow some temporal (stage to stage) comparison and measurement. The image data has its own digital coordinate system corresponding to the original section planes and discrete (voxel) coordinates which represent the basic resolution of the database. To accommodate both coordinate systems the image data is held in its original digital form and an affine transform (scaling, 3D rotation and translation) is defined which converts image coordinates to "real-space" embryo coordinates.

The real space embryo coordinates are defined by an origin and two spatial directions which define two of the axes of a right-handed Cartesian coordinate system. For the older embryos the origin is defined to be the rostral end of the notochord which also defines the direction of the z-axis. The x-axis is then aligned to point through the centre of the adjacent dorsal part of the neural tube. By this definition and at this location in the embryo, the x-axis points ventral-dorsal, the y-axis points right-left and the z-axis points caudal-rostral (figure 2).

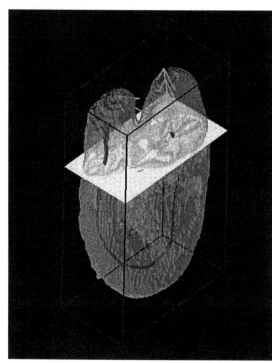

Figure 2: Atlas coordinate systems. The original digital coordinate frame is represented by the blue bounding box and is related to the original section images. The embryo coordinates are shown in yellow with origin at the rostral end of the notochord (orange). The x,y,z axes are shown with one, two and three arrow heads respectively. The outline of the embryo is shown as a transparent shell for orientation.

This definition will suffice for all embryos older than Theiler stage 11 when the notochord becomes visible. For earlier stages the origin and caudal direction can be defined by the primitive streak with symmetry used to define the other coordinates.

For blastocysts the shape of the blastoceolomic cavity provides the required origin and directions although the rotational symmetry implies that the orientation of the x- and y-axes is arbitrary. For each reconstruction, the rule used to determine the "real" coordinate frame is accessible from the database.

Atlas Temporal Coordinates

The time sequence of "significant" embryological events is not linear in the sense that the time resolution required to distinguish these events varies from fertilization to birth. Furthermore, it is well known that embryos of different strains develop at different rates. For this reason the use of "clock" time, for example days *post coitus* (dpc), is not always useful as well as being difficult to determine. To overcome this issue other time-scales have been defined which depend on developmental events, for example the first appearance of "fingers", or the closure of the eyes. The most widely used scheme for such *staging* is that due to Theiler [1,16]. For particular periods of development other staging systems are important namely Downs and Davies (D&D) [17] for the early embryos, and *somite count* for the period from about 8 to 10 days dpc.

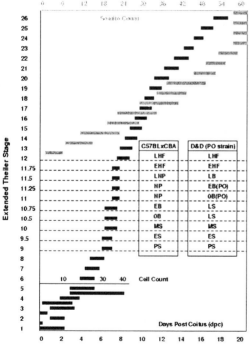

Figure 3: Edinburgh Mouse Atlas staging criteria and relationships shown in terms of the extended Theiler staging. Colour coded bars and scales are provided for the numerical measures. The relationships between the Downs & Davies (D&D) system for the two strains C56BLxCBA and PO shown in the boxes. Abbreviations are: LHF - late head fold, EHF - early head fold, LNP - late neural plate, NP - neural plate, EB - early (Allantoic) bud, OB - no bud, LS - late (primitive) streak, MS - mid-streak, ES - early streak, PS - pre-streak.

Experimentally determined gene-expression patterns will come from embryos that may not match the defining criteria for a particular stage because of inter-strain differences or simply natural variation and the embryos may not have been staged using the Theiler system. In this situation the data will be submitted with the authors' best estimate in whatever staging system used, and kept so that a user can be aware of how the stage was estimated. For search, that temporal information will be converted to an equivalent Theiler stage *range,* since in general there may not be sufficient precision to determine a single corresponding Theiler stage. The transformation between different staging systems is encapsulated in the table published by Bard *et al* (1998)[1] and is reproduced on the WWW pages of the Mouse Atlas Project (genex.hgu.mrc.ac.uk) and shown graphically in figure 3.

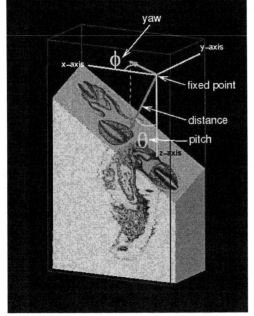

To accommodate the finer grain descriptions of the development sequence provided by Downs and Davies [17] for the early embryos, the "integer" Theiler stages have been extended to "floating-point" thereby defining sub-stages in the Theiler system.

Figure 4: Edinburgh Mouse Atlas section viewing coordinates. The fixed point can be redefined as required

Viewing Coordinates

To submit data to the database or to make queries involving spatial coordinates the user must be provided with an interface in which to define spatial position or region. The simplest interface appropriate to this task presents the user with a section view

---

[1] Submitted to *Development.*

which is a cut at an arbitrary orientation through the voxel model. The *viewing coordinates* define this section, how it is displayed on the screen and the transform from screen to model coordinates for mapping data or DB query.

To define the sectioning plane two angles are required to define the viewing direction and a perpendicular distance from a fixed point will define the section. For the viewing angles we define *pitch* as the angle by which the view is tilted away from the vertical, and *yaw* as the rotation of that view around the vertical. These two angles are the usual spherical coordinates, θ and ϕ and define the normal to the viewing plane. The other parameters that define the plane are the fixed point $\mathbf{f}$ and distance d.

The transformation from the image coordinate $\mathbf{r} = (x,y,z)^T$ to section coordinate $\mathbf{r}'$ $= (x',y',z')^T$ is given by

$$\mathbf{r}' = R\,(\mathbf{r} - \mathbf{f}), \text{ with } z' = d,$$

and the rotation matrix R is defined in terms of rotation by the three Eulerian angles ξ, η, ζ (xsi, eta, zeta) [18] by

$$R = R_\xi\, R_\eta\, R_\zeta, \text{ where}$$

$$R_\xi = \begin{pmatrix} \cos(\xi) & \sin(\xi) & 0 \\ -\sin(\xi) & \cos(\xi) & 0 \\ 0 & 0 & 1 \end{pmatrix},$$

$$R_\eta = \begin{pmatrix} \cos(\eta) & 0 & -\sin(\eta) \\ 0 & 1 & 0 \\ \sin(\eta) & 0 & \cos(\eta) \end{pmatrix} \text{ and}$$

$$R_\zeta = \begin{pmatrix} \cos(\zeta) & \sin(\zeta) & 0 \\ -\sin(\zeta) & \cos(\zeta) & 0 \\ 0 & 0 & 1 \end{pmatrix}.$$

The two Eulerian angles ξ and η are equal to pitch and yaw respectively and with the fixed point and distance determine the section plane.

To determine the transform from screen to voxel coordinates the third Euler angle must be fixed. The choice of this angle depends on how the user wishes to view the data and we define two viewing schemes termed the "up-is-up" and "statue" modes. For "up-is-up" the user defines an up-vector within the 3D reconstruction so that the projection of that vector onto the selected plane will be parallel to the displayed y-axis. By default this vector is defined to point up through the head so that a coronal or sagittal view will be displayed with the head neural tissue at the top of the image. For this mode the third Eulerian angle zeta is calculated by finding the component of the up vector **u** that is perpendicular to the viewing direction **v**, i.e.

$$\mathbf{w} = \mathbf{u} - \alpha\mathbf{v}, \quad \text{where } \mathbf{v} = R \begin{pmatrix} 0 \\ 0 \\ 1 \end{pmatrix}$$

and R is the rotation matrix for $(\xi, \eta, \zeta) = (\theta, \phi, 0)$. This vector is in the original coordinate frame and the required angle is determined by transforming w to the viewing coordinates and calculating the angle with respect to the y' axis:

$$\zeta = -\tan^{-1}\left(\frac{w'_x}{w'_y}\right) \quad \text{where } \mathbf{w'} = R\,\mathbf{w}.$$

The "statue" mode can be understood by considering the voxel image to be a statue with the user walking around on a horizontal plane i.e. the x-y plane of the image. The section seen by the user is rotated about its axis of intersection with the horizontal for display on the screen. For the user the effect is that rotating a section view around a vertical axis will result in a gradual rotation of the displayed image. The Euler angles for this viewing mode are $(\xi, \eta, \zeta) = (\theta, \phi, -\theta)$.

Navigation

Locating a specific section within a 3D volume can be quite difficult so, in addition to the option of panning though to a recognizable position which can then be defined as the fixed point, a number other navigational aids are provided. The first is the option of interactively defining a second fixed point which then reduces the number of degrees of freedom to one, namely rotation around the line joining the two points. This is termed the *torsion*, denoted τ, and uniquely defines the viewing direction

$$\mathbf{n} = \cos(\tau)\,\mathbf{n}_2 + \sin(\tau)\,\mathbf{n}_3,$$

where the orthonormal vectors $\mathbf{n}_1$ and $\mathbf{n}_2$ are defined by rotating the coordinate frame so that the z-axis is parallel to the vector joining the two fixed points, namely the vector

$$\mathbf{n}_1 = \frac{\mathbf{f}_2 - \mathbf{f}_1}{\left|\mathbf{f}_2 - \mathbf{f}_1\right|} \; .$$

This defines a viewing direction (θ_0, ϕ_0) and corresponding rotation matrix R_0 by $\cos(\theta_0) = \mathbf{n}_{1z}$ and $\tan(\phi_0) = \mathbf{n}_{1y}/\mathbf{n}_{1x}$ and hence

$$\mathbf{n}_2 = R_0 \begin{pmatrix} 1 \\ 0 \\ 0 \end{pmatrix} \text{ and } \mathbf{n}_3 = R_0 \begin{pmatrix} 0 \\ 1 \\ 0 \end{pmatrix} \; .$$

The new viewing angles (θ, ϕ) are then given by $\cos(\theta) = \mathbf{n}_z$ and $\tan(\phi) = \mathbf{n}_y/\mathbf{n}_x$, and are parametrised by τ, which ranges from 0 to 2π.

The second means of navigation is to use *fiducial* points, i.e. points that have been previously defined and exist in the Atlas database for the purposes of navigation. Selecting a single point defines a rotation center, two points will define a line as above and a third will fully define a plane.

Other navigational support is provided by a 3D feedback window (see figure 5) in which the section plane is displayed relative to the image bounding box and selected anatomical structures.

Data Submission

The Edinburgh Mouse Atlas has been developed primarily to support a gene-expression database by providing a spatial framework for the mapping expression and anatomical information. The methods for mapping spatially organized data onto the reconstructions can be categorized under the headings: text, painting or warping.

Text submission:

The spatial domain is described by listing anatomical components or existing database entries. Typically the user will define a domain by selecting anatomical components with the assumption the required domain is the *union* of the component domains (which by definition do not intersect). If in addition the user wishes to use predefined gene-expression domains these can be used to refine the new domain by using the set operations *union*, *intersection* and *difference* on the corresponding sets of voxel locations.

138

Painting:

An alternative mean of defining the domain is for the user to delineate or paint the new domain directly onto the reconstruction using the Edinburgh Mouse Atlas paint program (MAPaint see figure 5) or equivalent software [9]. This allows the user to select an arbitrary section and delineate regions directly using a variety of painting "tools" [9]. This program also allows the editing of domains defined by any other means therefore can be used in conjunction with a text definition of the domain to allow additional refinement. This is likely to be more efficient than defining the whole 3D domain *de novo*.

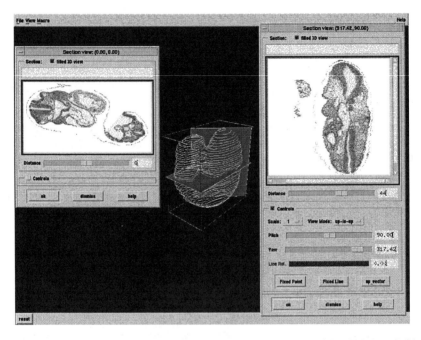

Figure 5: The main feedback window and two section views of the Edinburgh Mouse Atlas Theiler stage 14 embryo in the program MAPaint showing the 3D feedback and view controls.

Warping:

Submission using text and/or painting can be time consuming and most graphical data will be entered by using image processing both to extract the required pattern and to transform or *warp* the data onto the atlas. Before generating the submission it is assumed that the user has digitized the required experimental data, then the simplest means is to locate the matching section from the atlas and for the user to define "tie-points" that can be used to produce a warp transformation, e.g. as a thin-plate spline [19] or polynomial [20], to map the experimental data onto the atlas image. If the resultant transformation is not correct then the user can add additional tie-points or directly edit the resulting domain using painting. There are many possibilities for determining the warp transform including a fully automatic matching [8]. In practice the mechanism must be fast enough to allow interactive adjustment and editing.

So far we have considered section data which assumes that the experimental section corresponds to a planar section through the atlas embryo. If the experimental data is in the form of a 3D image then full 3D warping will be necessary e.g. [13].

Data Extensions

The initial implementation of the image-mapped database will store the domains (binary image) of gene-expression only. This means that expression strength information can only be encoded as separate domains, e.g. weak, medium and strong, or by storing a dithered version of the grey-level image. The Woolz image format includes a simple extension to grey-level types within the image domain and therefore we plan to extend the schema to allow full grey-level images of gene expression data to be submitted. This means that the query language can be extended to use grey-level operations such as gradient and intensity.

References

1. Theiler, K. *The House Mouse: Atlas of Embryonic Development*, (Springer-Verlag, 1989).
2. Baldock, R A, Bard J, Kaufman, M H & Davidson D, *A Real Mouse for your Computer*, BioEssays **14**(1992)501-502.
3. Ringwald M, Baldock R, Bard J, Kaufman M, Eppig J, Richardson J E, Nadeau J & Davidson D, *A Database for Mouse Development*, Science **265**(1994) 2033-2034.
4. Davidson D, Baldock R, Bard J, Kaufman M, Richardson J E, Eppig J T & Ringwald M, *Gene-Expression Databases*, *In Situ* Hybridisation: A Practical Approach, ed. Wilkinson D, (IRL press, Oxford, 1998).
5. Davidson D, Bard J B L, Brune R M, Burger A, Dubreuil C, Kaufman M, Quinn J, Stark M & Baldock R A, *The Mouse Atlas and Graphical Gene Expression Database*, Sem. Cell & Dev. Biol. **8**(1997)509-517.
6. Eppig J T & Richardson J E, *MGD and GXD*, this issue.
7. Guest E & Baldock R A, *Automatic Reconstruction of Serial Sections Using the Finite Element Method*, Bioimaging **3**(1995)154-167.
8. Baldock R A, Verbeek F J & Vonesch J-L, *3-D Reconstruction for Graphical Databases of Gene-Expression*, Sem. Cell & Dev. Biol. **8**(1997)499-507.

9. Arya M, Cody W, Faloutsos C, Richardson J & Toga A, *A 3D Medical Image Database Management System,* Comp. Medical Imaging and Graphics, **20**(1996)269-284.

10. Hohne K H, Pflesser B, Pommert A, Riemer M, Schiemann Th, Schubert R & Tiede U, *A new representation of knowledge concerning human anatomy and function,* Nature Medicine **1**(1995)506-510.

11. Toh M Y, Ackerman M, Rodgers R P & Banvard R A, *Visible Human Project on the Internet - World-Wide-Web Access,* Radiology **201**(1996)9501.

12. Williams B S & Doyle M D, *An Internet Atlas of Mouse Development,* Comp. Medical Imaging and Graphics **20**(1996)433-447.

13. Thompson P & Toga A W, *A Surface-Based Technique for Warping 3-Dimensional Images of the Brain,* IEEE Trans on Medical Imaging **15**(1996)402-417.

14. Piper J & Rutovitz D, *Data Structures for Image Processing in a C Language and Unix Environment,* Pattern Recognition Letters **3**(1985)119-129.

15. Booch G, *Object-Oriented Analysis and Design with Applications,* 2nd edition (Benjamin Cummings, California, 1994).

16. Kaufman M H, *The Atlas of Mouse Development* (Academic Press, London, 1992).

17. Downes K M & Davies T, *Staging of Gastrulating Mouse Embryos by Morphological Landmarks in the Dissecting Microscope,* Development **118**(1993)1255-1266.

18. Whittaker E T, *A Treatise on the Analytical Dynamics of Particles and Rigid Bodies,* 3rd edition (Cambridge University Press, Cambridge UK, 1927).

19. Bookstein F L, *Thin Plate Splines and the Atlas Problem for Biomedical Images,* Lecture Notes in Computer Science **511**(1991)326-342.

20. Olivo J-C, Izpisue-Belmonte J-C, Tickle C, Boulin C & Duboule D, *Reconstruction from Serial Sections: a tool for developmental biology: application to hox genes expression in chick wing buds,* Bioimaging **1**(1993)151-158

12 FLYBASE: GENOMIC AND POST-GENOMIC VIEWPOINTS

The FlyBase Consortium[1]

FlyBase, The Biological Laboratories, Harvard University, 16 Divinity Avenue, Cambridge, MA 02138

Introduction

FlyBase is a genomic and genetic database of the family Drosophilidae. The vast majority of data concern the premier genetic model among these flies – *Drosophila melanogaster*. *D. melanogaster* has been at the leading edge of investigations of many aspects of genome structure, gene regulation, intercellular communication, developmental genetics, neurogenetics, cell biology and population genetics. FlyBase has been developed to represent the core information regarding these studies.

FlyBase is currently undergoing a major reorganization and expansion. The original FlyBase consortium, which focused on capture and presentation of the Drosophila genetic/genomic "literature", has now merged efforts with the public informatics groups of the two major Drosophila Genome projects (Berkeley Drosophila Genome Project – BDGP, and European Drosophila Genome Project – EDGP). While separate servers still exist for FlyBase, BDGP and EDGP, this is a transitory condition. Within the next year, the goal is to have the combined data of these projects homogenized and presented on a single integrated server with several mirror sites around the world. Having a unified genome project and literature database is clearly of benefit to the community; it of course represents considerable challenges to FlyBase in the integration and interconnection of these broad ranges of data types. As part of the integration process, it is likely that the content and organization of reports will undergo considerable evolution. Because it is being written as the FlyBase transition is occurring, this report will focus on the current views of FlyBase and how these are expected to change as a result of the integration

---

[1] Corresponding Author: Bill Gelbart, gelbart@morgan.harvard.edu.

process. Because the databases are currently in heterogeneous format, the road to integration will be an interesting experiment in database interconnectivity.

The FlyBase Consortium Model

FlyBase has opted for a distributed structure, in which several remote sites participate in producing a single integrated database. With the inclusion of BDGP and EDGP, the FlyBase Consortium includes 5 groups, located at: Harvard University, University of Cambridge, Indiana University, University of California – Berkeley, and the European Bioinformatics Institute. The advantages of this approach are the small sizes of the individual groups, the inclusion of several project directors who in part act as a set of internal advisors, and the heterogeneity of viewpoints that are represented. FlyBase has concluded that this distributed organization works extremely well for its purposes. It does however place considerable emphasis on issues of data transfer, synchrony, and integration.

The Subdivisions of Data Responsibilities within FlyBase

The data outputs of the BDGP and EDGP projects are their individual responsibilities, as are the literature data outputs of the original FlyBase group. These include:

BDGP:
1. P1/BAC STS content physical map of the genome.
2. Genomic sequence of the autosomes.
3. Characterization of ESTs from a variety of cDNA libraries.
4. Sequencing of representative near full-length cDNAs.
5. Autosomal P element insertion mapping and characterization.
6. Computational identification of genes.
7. Incorporation of the above data into genome level maps.

EDGP:
8. Cosmid/BAC physical map of the genome.
9. Genomic sequence of the X chromosome.
10. X chromosomal P element insertion mapping and characterization.
11. Computational identification of genes.
12. Incorporation of the above data into genome level maps.

Literature:
- Bibliographic information, including links to bibliographic databases such as PubMed and BIOSIS.

- Genes, including links via published sequence and functional similarities to other molecular and community databases.
- Gene products: transcripts and proteins, including structural and expression pattern data.
- Integrated gene order maps, incorporating recombinational, cytogenetic and molecular information.
- Annotated molecular maps: reference sequence gene maps; regional physical maps.
- Alleles: wild-type, mutant and engineered.
- Chromosomal aberrations.
- Engineered and natural transposons and their insertions in the genome.
- Fly strains: principally, the publicly-funded stock collections.
- Contact information for Drosophila researchers.

FlyBase "literature" is meant broadly, and includes hard copy publications, sequence databank entries, bulk submitted data, stock lists, textual personal communications, etc. The principle is that all information in FlyBase is attributed, and is linked to a hard copy or electronic text. The literature database consists of records containing a mixture of controlled or structured fields and free text descriptions to extend the structured information. Extensive internal and external cross-referencing of related data objects is included. For these controlled fields, FlyBase has developed extensive controlled vocabularies, e.g., for phenotype, anatomy, mutagen, function of gene product.

Data Curation Approaches

The data of the BDGP and EDGP consist almost exclusively of structured outputs of high throughput genomic analyses. There is considerable curational input at such levels as gene predictions in genomic DNA sequences, and genetic analyses of P element insertions.

The literature curation aspect of FlyBase began in 1992. It took as a starting point the compilation of Lindsley and Zimm [1], which was largely current to the beginning of 1990. The goal was to capture information from the post-1989 primary literature, and selectively curate earlier material as deemed necessary. Curation of genetic information occurs shortly after journal publication for those journals considered to be the major ones used by the Drosophila research community. For papers with molecular information (e.g., on gene structure, transcripts, proteins, expression patterns, transposons or their insertions), these receive a second round of curation with sets of papers relating to the same gene being curated together; available GenBank/EMBL/DDBJ records are examined at the same time, since a better picture of these molecular data classes emerges from simultaneous consideration of a group of related papers. Curated reference annotated sequence records have recently been added to molecular curation (further discussed in Sequence Annotation, below).

Creation of External Database Links

FlyBase receives daily updates of GenBank/EMBL/DDBJ records of the family Drosophilidae, and captures the links to valid FlyBase gene symbols and identifers. Tables of these links are shared with the sequence databanks. Similar procedures are used with regard to links to other sequence databases, particularly SwissProt. Links from the literature to "homologs" in other species are selective, and are based upon stated sequence similarities in papers. Based on these statements, FlyBase captures the valid gene symbol and identifier for the declared "homolog" in the foreign community database. The BDGP and EDGP use sequence similarities as one aspect of gene prediction, and capture and maintain links to the strongest BLAST similarities in GenBank/EMBL/DDBJ, focusing on the major genetic systems wherever possible.

Data Coordination within FlyBase

It will be obvious from the above that there are many data objects in FlyBase that are common to the genome projects and the literature (the latter being the product of the entire Drosophila research community). In order to bring these data into a single structure, we will need to integrate and homogenize the data in a stepwise process.

All FlyBase data will move, as a first step into either of two databases: one being an integrated Genome Project database and the other an integrated Literature database. (The integrated Literature database is already in production use; the integrated Genome Project database will be implemented soon.) At this step, considerable data validation and homogenization occurs. The next step will be to interrelate and homogenize literature-derived and genome project data of the same class. This has been done successfully as an experiment for selected data classes, such as transposon insertions. Based on this experience, expert annotators will need to examine the data to ensure that identical objects with variant names are being recognized as identical, and valid symbols will need to be agreed upon and propagated to all of the relevant working and intermediary databases. Based on FlyBase's experience with the integrated Literature database, procedures will be established such that new data objects at each site will receive valid symbols and interconnections to other valid objects as they are introduced into the database. The final step will be to map field identities between the integrated Literature and integrated Genome Project databases and thereby permit the data to be housed in one structure.

Public Access to FlyBase Data

The current home WWW servers for genome project and literature data are:
- BDGP: http://fruitfly.bdgp.berkeley.edu/
- EDGP: http://edgp.ebi.ac.uk/
- FlyBase literature: http://flybase.bio.indiana.edu/

The current organizations of these sites are different, reflecting the ways in which the developers separately envisioned suitable presentation formats for their principal data objects. These will evolve into a common presentation format for the integrated server. On the integrated server, genome project and community data will be presented in merged reports, or in genome project-only format, according to the preferences of the user. In developing the integrated server, some of the important considerations will be:
- graphical overviews summarizing large data sets.
- use of publicly available, nonproprietary software tools to facilitate mirroring.
- provision of user-selected formatting options.

A challenge for all database projects is how to cope with the ever-increasing acceleration in rates of data capture. FlyBase's approach to this issue is to condense much of the information into summary graphics wherever possible. This is currently under development for two areas.

One major mode of access to FlyBase data is via chromosomal location. For chromosomal or genomic information, a series of Java-based map displays have been developed by the BDGP and Neomorphic, Inc., and these will be incorporated into the integrated FlyBase server. These displays may be viewed at the BDGP WWW server. These dynamic displays present genomic and genetic data in several levels, from a "low magnification" view based on the polytene chromosome map, all the way through annotated maps of sequenced genomic clones and sequence alignments.

Another major route of access is through phenotype or expression pattern. Here, conventions for naming anatomical parts are not as well established as for chromosomal location. To facilitate user access to phenotype or gene expression pattern information, FlyBase is developing dynamic anatomical Java displays, in which anatomical drawings can be selected so that the name of the relevant structure is displayed, and will enable queries for genes whose phenotypes affect this structure, or whose transcripts or proteins are expressed in this structure. A sample of such a display is shown in the images section of the FlyBase server.

Other considerations relate to providing flexible and effective access to FlyBase data. Because of variability in the reliability and speed of Internet connections, particularly between continents, FlyBase has benefited from the willingness of regional sites to serve as FlyBase mirrors. Currently, sites exist in the United States, England, France, Israel, Japan and Australia. Typically, these mirrors are updated nightly. Users are encouraged to connect to local sites for the most rapid response time. To facilitate development and maintenance of mirrors, the FlyBase WWW server has developed a series of portability features, including the use of readily mirrored, nonproprietary software, nightly updating of the mirror sites, and customized user-selected format preferences. For example, the user preferences would allow an individual (or an entire mirror site) to connect to the nearest server for external database links.

Public Server Update Schedule

The FlyBase data classes are updated asynchronously, and this is expected to continue indefinitely. Considerable editorial work occurs in the batch processing of records, and for this reason FlyBase has opted against a daily update procedure. Sequence records need to be the most timely; for this reason, the genome databases are updated weekly. These updates include the GenBank/EMBL/DDBJ accessions; it should be noted that all available genomic sequences are posted immediately to the HTG (high throughput genome) section of the sequence databanks. Public bibliographic, genetic and molecular data classes are updated monthly, but on independent schedules.

Genetic nomenclature is in constant flux, sometimes because a group of laboratories have not yet reached consensus on an appropriate gene name, but other times because research on seemingly independent genes converge when genetic or molecular analyses reveals that they are working on the same entity. Because a gene symbol is part of the name of many related entities -- alleles, transcripts, proteins, transposons, insertions, etc. – the updating process requires substantial validation, editing and conflict resolution. In this process, there is significant contact with authors on specific issues.

Future Issues and Concerns

Sequence Representations in FlyBase

While the first genomic sequence for *Drosophila melanogaster* will be completed by the end of 2001, this will in many ways just be the beginning in terms of thorough genomic analysis of the fly. Thus, there is a need for continual and indefinite maintenance of the computed and experiment-based annotation of this sequence. Indeed, it is through such annotation that the genomic sequence becomes meaningful. Thus, much of FlyBase can be viewed as statements attached directly or indirectly to the sequence-level annotation features.

There are two classes of annotation maintained by FlyBase. The first of these is clone-based, in which the deposited fully-sequenced BDGP and EDGP clone records are periodically re-analyzed for computational and experimental predictions of transcription units and coding sequences (gene calls). The information used in these analyses will be restricted to information produced by the genome projects themselves, together with information derived from existing GenBank/EMBL/DDBJ accessions.

The other class of annotation is directed at capturing community-based annotation features. There is a very active Drosophila research community that produces connections between genes as molecular entities and genes as modulators of phenotypes. This is true not only at the whole-gene level, but also in terms of individual features of the gene -- introns and exons, coding sequences, enhancers and silencers, boundary domains, mutant alterations, transgenic rescue fragments, etc. While the community data set is rich, much of it has not been incorporated into GenBank/EMBL/DDBJ records, either because the data are only known at the restriction map level or by omission. As part of the literature molecular curation, FlyBase is capturing community annotation features, not only from the GenBank/EMBL/DDBJ records themselves, but also from the primary literature. Interactive tools have been developed that attach these features to a reference gene sequence – typically the relevant finished genome project sequence for that portion of the genome. All features are given unique FlyBase symbols and identifiers, so that biological information about these features can be directly connected to them. Many features defined only at the restriction fragment level can be connected to the sequence level map through comparisons of the published and computed restriction maps, and the identification of landmarks common to the publications and the reference sequence record.

The final step in the compilation of reference sequence records is to merge the computed and literature annotations into a combined virtual sequence of the *Drosophila melanogaster* genome. These sequences will represent the best summation of FlyBase's understanding of the structure of the genome, its genes and their products. Because the different aspects of sequence level curation are the responsibilities of sites spread over two continents, the process of integration of these data will require curators to communicate via editable real-time graphical displays.

The current plan is that once these two classes of annotation are created, they will be updated on a 6 month cycle. Once the entire genome is complete, this will require maintenance of records spanning approximately 600 kb per day of computational gene prediction of clones. As only a small subset of Drosophila genes currently are being studied at the molecular level by the community (although this may change dramatically once the entire genome sequence is available), the rate of steady-state sequence updates with regard to community annotations is impossible to estimate. Clearly, direct user involvement will be of great benefit in scaling this community annotation effort, and might occur via more systematic use of sequence databank feature tables and/or via FlyBase direct user submission tools. Scaling this effort and obtaining effective community involvement in data capture is an important challenge for the future.

Maintaining Extensive Hyperlinks between Community Databases

One of the outstanding features of the WWW is the ability to embed cross-links between related items on different servers. FlyBase maintains extensive tables of connections to support such cross-links, especially to sequence and community databases. We recognize, however, that this effort is incomplete and not sustainable by FlyBase alone. For example, FlyBase reflects statements of "homology" in the primary literature. Except for fully sequenced organisms, however, these declarations of "homology" are generally based on incomplete information, and more similar members of the same protein family may well be identified subsequent to publication. Further, FlyBase considers papers that involve the identification of a Drosophila gene with sequence similarity to a gene in another taxon as being in its literature curation domain, but the converse is not true. Papers in which a Drosophila sequence is used as a probe to identify related sequences in another organism are viewed as appropriate for curation by the community database for that organism, if it exists. There is clearly a need for an independent database of links. at least between the genomes of the major genetic systems, that these organisms' community databases contribute to and use for more systematic interconnections.

Genome, Proteome and Phenome Views of FlyBase

Just as the availability of full genomic sequence has changed the way research is done, functional genomics – that is, the high throughput analysis of gene products – will clearly have a similar, if not more dramatic, impact. One tremendous advantage of genomic information is the increased efficiency of positional cloning and gene identification. Once the gene is identified, however, researchers are usually interested in understanding relationships at the level of gene products and their physical, cellular, developmental or behavioral interactions. Much modern research exploits the major genetic systems to answer sophisticated questions about the relationship of the "proteome" (the entire constellation of protein products produced by a genome) with the "phenome" (the constellation of phenotypes controlled by the various genetic units of the genome).

We are already seeing this trend in Drosophila research. Much research focuses on fleshing out pathways through a combination of physical methods to identify protein-protein, protein-RNA or protein-DNA interactions, together with genetic screens to identify direct or indirect interactions based on phenotypes or gene expression patterns. It is thus important for FlyBase to recognize and support data representations and reports based on relationships among gene products in addition to those relationships based on chromosomal location. Some ways of addressing this need can be addressed now; others present substantial technical hurdles.

The FlyBase architecture supports the curation of different versions of a gene product –RNAs or polypeptides or molecular complexes – as different data objects, so that annotations can be attached to the appropriate objects. This is an essential part of an organism-specific data model, since much of the regulation of cellular function boils down to gene products that can be toggled between alternative states based on allosteric interactions, subunit modifications, or differential subunit interactions.

Describing the interactions and the pathways is an even larger and more difficult task. Much of the available physical interaction data involves in vitro assays, usually in heterologous systems. These data are often hints or suggestions of *possible* interactions rather than readily verifiable ones. Genetic interaction data have their own set of pitfalls. While the individual observations can be represented, our ability to compile them into computed pathways is impaired by the inherent limitations of the current data sets. Thus, we need to capture and represent data in a manner that reflects the current state of knowledge, but that will be of value once better standards and methods are available. This represents a considerable challenge at the strategic and computational levels.

Another aspect of the problem are those of spatial pattern: descriptions of anatomical phenotypes and gene expression patterns. Were rigorous representations of spatial pattern possible, these could be used in combination with interaction data to distinguish among possible interactions. (For example, two proteins that are shown to physically interact but which are never expressed in the same tissues are unlikely to interact in a biologically meaningful way). FlyBase has developed an extensive ontology of anatomical parts, and using this vocabulary, phenotypes and expression patterns are captured. Either the authors or the curators, however, end up throwing away a great deal of data in turning two or three dimensional spatial information into text. Similarly, dependence on text terms to support user queries places inherent limitations on the depth of questions that can be answered. Ultimately, it will be important for tools to be developed that can effectively capture quantitative spatial information. Only in this way can these data can be directly queried without imposing a strong filter on the data set through its conversion into much coarser textual objects. This is obviously a major long term issue which is already receiving attention, and we can expect that it will continue to be an important area for computational research.

References

1. Lindsley, D. and G. Zimm. *The Genome of Drosophila melanogaster.* Academic Press, NY., 1992, 1133 pp.

13 MAIZEDB: THE MAIZE GENOME DATABASE

Mary Polacco and Ed Coe

USDA-ARS Plant Genetics Research Unit
Department of Agronomy 210 Curtis Hall
University of Missouri -- Columbia
Columbia, MO 65211

INTRODUCTION

In 1923, R. A. Emerson addressed a detailed letter to "Students of Corn Genetics', soliciting community solutions to issues of gene nomenclature. (Emerson 1923) Building on this stimulus, Emerson hosted a 'cornfab' in his hotel room, during the December 1928 Genetics meetings in NY. April 1929, Emerson and colleagues disseminated the first Maize Newsletter (MNL): a mimeographed summary of the 'cornfab', along with a list of 20 available stocks, a list of curators for individual linkage groups, a summary of available linkage information, and 78 references where linkage data were to be found. The first published linkage map compilation in 1935 showed 62 loci. (Emerson et al 1935) In 1991, USDA-ARS initiated a plant genome database project, and tasked the editor of the MNL, Ed Coe, to develop a maize genome database. The 1991 MNL, now volume 65, included 1439 Stocks available from a USDA-ARS funded Stock Center, 840 entries on the Gene List with 423 key references, an additional 776 references from the annual literature and some 950 colleague addresses. Data from this issue were transferred into the Fall 1991 prototype MaizeDB.

The 1991 prototype was based on insight gained by the Coli Genetics Stock Center, New Haven, CT with an industry standard software, Sybase, for the database management system, and the development of Genera software (Letovsky & Berlyn 1994) to create forms for query and data entry and define entities without writing code. This choice of software has permitted concentrating resources on data curation by a team of 'biologists', with support of a systems analyst. The Sybase software had been employed by various other species genome databases, including the human genome. A sample of the Genera form specification for the short locus query form is appended to this chapter, along with the form specified, and a sample query returned. Indexed flat-files which provide full-text searching have been a standard feature of the Genera software. In 1994, MaizeDB, http://www.agron.missouri.edu employed Genera upgraded for WWW form and full-text query access; the upgrade also permitted facile linking with external databases. Data relevant to the maize genome exists in major sequence, reference and germplasm databases, as well as other species-specific genome databases, including *E. coli*, yeast and other plant genomes. Reciprocal, record-to-record links with SwissProt were established June 1994, and followed soon thereafter with reciprocal links to GRIN, the Germplasm Resource Information Network for the US.

DATA CONTENT

MaizeDB maintains the current genetic maps for the community, together with supporting documentation, and information about gene function and expression. Supporting documentation may includes(1) data analyses summaries, such as QTL experiments; (2) raw data, map scores and recombination data; (3) references, with address information for many of the authors; (4) access to research tools, including genetic stocks, DNA clones and PCR primer sequences. Gene function and expression are provided in comments on the locus pages, in the gene product, mutant phenotype, trait information about mapped and unmapped. Gene function and expression are provided in comments on the locus records, by links to gene product, mutant phenotype, and agronomic traits.

Over 110,000 records are currently maintained by MaizeDB; major classes of information are summarized in the below table. There may be upwards of some 40,000 genes in maize. Of the 15,764 maize loci in the database, there are 6,028 genes, where 1,188 have been mapped to chromosome arm or better by mutant phenotype or single copy probe. The MNL Gene List, now includes 1,512 unique genetic factors of which 827 have some map position. Loci have expanded from the 62 phenotypically identified genes on the 1935 map compilation to include pseudogenes, probed sites (PCR or RFLP); restriction fragments (mitochondrial maps), chromosomal segments, points and QTL (quantitative trait loci).

Jan 94	Jul 98	Entity
5,032	16,121	Locus
21,401	21,834	Variation
2,055	10,633	Stock
94	617	Map
1428	7206	Map Scores
143	1730	Recombination Data
213	773	Phenotype
931	8032	Probe
1532	1585	Gel Pattern
1		Clone Library
16	529	Restriction Enzyme or PCR Primers: PCR primers
238	1061	Gene Product
0	46	QTL Experiment
0	201	QTL Linkage Analysis
0	199	Trait Evaluation Summary
0	67	Environment
3932	15,828	References
	1834	MNL articles
5394	16,222	Person
	4255	With address
58	490	Trait

The bins map strategy has been employed by the MNL editor to unify the extensive maps in the community, and without misrepresenting known order. Order of loci within a bin may be ascertained by examining the empirically determined maps, which are provided within the database.

The MaizeDB server also hosts the electronic MNL, abstracts for the annual meetings, and provides a public clearing house for gene nomenclature and registry of new names.

DATABASE DESIGN AND FEATURES.

The database contains 27 entities, 340 tables, 53 views, 50 triggers and 221 stored procedures. Stored procedures written at Missouri generate an ACeDB product that was first released in June 1993. Major enhancements to the 1991 prototype were implemented March 1993; QTL experiment representations were added 1994. Minor design upgrades, or example merging two entities complete with data, creating a Trait entity, and writing triggers have been accomplished with the core MaizeDB staff at Missouri. Schema may be found at our public ftp server (teosinte.agron.missouri.edu); both computer and biological documentation about fields, tables and entities may be accessed by WWW form query.

The database supports controlled vocabularies, with a thesaurus. All major entities have synonym tables. The most carefully curated terms are the entity "Type" terms, those defining relationships between loci, and the keywords used for reference annotation. The data entry forms permit selecting from the controlled vocabulary, and also entry of new terms if appropriate. In contrast to many of the controlled vocabularies, for example body parts, terms for entity types are kept to a minimum, and the use of broader and narrower terms avoided. Instead, there are tables for entity properties, where a locus of type Gene or Cytological Structure may share a common property, for example MNL Gene List.

EXTERNAL DATABASE LINKS

Over 13,500 records have over 62,000 links to some 30 external databases. When more than one record links to the same external database record, each link is counted in the summation. Links to loci are most often indirect, and linked directly to a gene product, probe, variation or reference. While only 2% of the loci may be linked to an external database, 10% of the locus variations have links, as do 27% of Probes, 32 % of Gene Products, 30% of the Stocks and 18% of the References, excluding the MNL and maize meeting abstract links.

Links to external databases are provided by two modes: user request for data from the other database, where typically a particular database may be selected for retrieval of a sequence, or by MaizeDB pre-determined choice or 'jump'. In both cases, external databases are treated as Persons. The code for the 'jump' is embedded in Genera and utilized in MaizeDB both for Person and for Loci; see also the genera specification for the locus query page in Appendix 1.

An example of user-choice for external database:

DB Key	DB	Variation
M61191	DDBJ	wx1-B6
M61191	EMBL	wx1-B6
M61191	Entrez	wx1-B6
M61191	GenoBase	wx1-B6
M61191	GSDB	wx1-B6

An example of a jump is provided by the maize locus wx1, which lists several related loci as below:

Relation	Locus
Left Marker	w11 white11
Right Marker	d3 dwarf plant3
Contained in	Dp9 Duplication 9
orthologous, putative	xWx Triticum aestivum
orthologous, putative	WX Oryza sativa

Clicking on one of the orthologous rice or wheat loci retrieves the Plant Genome database for that record. In contrast

DATA ENTRY

Form entry is provided by the Genera software utilized for non-WWW access to the database. A small group of both on-site and off-site curators have access to the central Sybase tables using this software. Other data is entered using customized scripts, developed by curators for large electronic notebooks in the community, or standard script, such as that developed in 1993 for journal article import from standard reference manager format. It has been upgraded by MaizeDB staff to enter books and book chapters, and also MNL articles. The reference loading software matches to previously entered references, fills in missing information and reports actions taken and ambiguities.

DISSEMINATION:

WWW form permit queries based on multiple attributes, and return lists of objects that match the selected constraints. A sample form for Locus is appended. Full text query offers an opportunity to the novice user to obtain a sense for how the data are represented. Flat-files for the full-text searching are computed monthly, but retrieve real-time data..

Special Data Formats:

In addition to form queries, based on selection and/or 'fill-in-the-blanks' query constraints, MaizeDB supplies several browser oriented formats on the WWW. These formats may be on-the-fly, with some or no user constraint options or they may be periodically extracted, with or without curator intervention. Hyper-links on lists will retrieve the current record from the database, based on the accession ID# in MaizeDB for the entity. Some of the lists are computed in real-time, others extracted periodically, either automated, the Core Marker list), or with curator intervention (the MNL Genelist).

1. Computed in real-time -- no user constraints.

Clicking at the Illionois Sotck Center page will initiate a real-teim query for the current listing of Stocks, either total or the 1998 additions. The catalog returned is extracted from the Stock and the Stock#Description tables and has hypertext links to the Stock entity. It provides the Stock Center accession, and a descriptive name of the Stock, as computed daily from a list of the variations associated with the Stock.

ACC	DESCRIPTIVE NAME
301B	bif2-N2354
302AA	d1-N446
302AB	d1-N339
306F	ref1-MS1185
307A	Sdw2-N1991
309A	a1-m3::Ds Sh2
309B	a1-m1-5718::dSpm
309C	a1-m1-5719A1::dSpm
309D	a1-m1-5719A1::dSpm; Mod Pr1

2. Real-time --with user constraints.

Example 1.

Map Scores by the BIN: user selects (1) the bin range, (2) the Mapping Panel of Stocks, and (3) format of product returned (hypertext, or comma-delimited). Data for this table are read from the Locus#Coordinates, and MapScores tables.

Map Scores for loci between bins 1.01 and 1.12 using Mapping Panel 57244

Coord	Bin	Locus*	Map Score*
11	1.01	csu738	AHBHBCHHBBHHHAHHHBAHBHHAHH-BHAHHBABAABH-BBAABBCAHHHBBB
11	1.01	tub1	AHBHBBHHBBHHHAHHHBAHBHHAHHBBHAH HBABAAB-HBBAABBHAHHHBBB
11.9	1.01	umc94a	AHBHBBHHBBHHHAHHHBHHBHHAHHBBHAH

			HBABAABHHBBAABBHAHHHBBB
12	1.01	csu589	AHBHBBHHBB- HHAHHHBHHBHHAHHBBHAHHBAHAABHHBB AABBH-HHH-BB

* hyper-text linked to MaizeDB entity

Example 2.

Formatted Person address. Constraint options are the city and or the person's last name. Data may be returned in a choice of 3 formats: (a) address only, (b) address and phone, as required for FedEX, or with (c) address, Email, phone and fax. Data for the below example are read from the Person, Person#PhoneNos and Persone#EMailAdresses tables.

City="Urbana", and format c, above; only one of the 18 addressed returned is shown:

> Marty Sachs
> USDA/ARS
> S108 Turner Hall
> 1102 S. Goodwin Ave
> Urbana IL 61801
>
> (217)244-0864/333-9743lab
> (217)333-6064 (fax)
> msachs@uiuc.edu
> (verified May 7 1998)

3. Computed periodically, no curator intervention.

Table of Core Marker information is automatically computed each week. The table content was defined by the UMC RFLP laboratory. Core markers are loci that define the edges of the bins on the consensus map; this map is curated by Ed Coe, at Missouri, with MaizeDB staff assistance.

Data are read into the output table from the Locus#Coordinates, Probe and Probe#Comments tables, as restricted by the a property 'Core Marker', stored in the Locus#Properties and Probe#Properties tables as two independent controlled vocabularies.

Sample rows:

Probe*	Locus*	Map*	Bin	Type	Insert (bp)	Enzyme
p-tub1	tub1	1	1.01	genomic	158	EcoRI/H indIII
p-umc157	umc157(chn)	1	1.02	genomic	1220	PstI
p-umc76	umc76(gne)	1	1.03	genomic	760	PstI
p-asg45	asg45(ptk)	1	1.04	genomic	350	PstI
p-csu3	csu3	1	1.05	cDNA	1200	EcoRI /XhoI

*hypertext-linked to MaizeDB entity.

4. Periodically computed,-- coordinated with curator action.

The printed 1993 MNL Gene List (volume 67) was the first Gene List extracted from the database. The Gene List is a complex table, with the symbol, the bin location, the full name, a brief comment, putative or confirmed gene products and key references for selected loci. Hypertext links are provided to the MaizeDB Locus, Gene Product and Reference entities. Curatorial review: (1) checks that all appropriate loci are included, by examining the list of excluded maize loci of type="Gene" where there is no '*' associated with the name; (2) ascertains that the full name, the brief comment, any gene product(s) and appropriate references are updated; (3) monitors comments entered into the database for errors, completeness and currency. Creating the Gene List requires reading data into 3 de novo tables (genes, geneP, MNLGeneRefs) from several MaizeDB tables (Locus, Locus#Coordinates, Locus#Comments, Locus#GeneProducts, Anything#References, References). The only constraints are the Locus Property, MNL GeneList and the reference Annotations, First Report or Gene List.

Two sample rows.:

bt1, bin(s) 5.04, *brittle endosperm1*, mature kernel collapsed, angular, often translucent and brittle (alleles sh3, sh5), may encode amyloplast adenylate translocator ref: 466, 867

pl1, bin(s) 6.04, *purple plant1*, Pl1 plant tissues have light-independent pigment, pl1 blue light-dependent; Pl1-Bh1, colored patches in c1 aleurone and in plant; transcriptional activator for flavonoid genes; SSR phi031, nc009, nc010 ref: 215, 216.

DISSEMINATION TO EXTERNAL DATABASES

Linking information is provided to SwissProt, to GRIN, and to EMBL, and to the Entrez-Genome division associated with GenBank. Mapping coordinates are placed into our public ftp file, and the Entrez-Genome division is notified when there is an update. The AceDB format for the database is submitted to the central server for Plant Genome Databases at the National Agricultural library.

FUTURE DIRECTIONS

The Plant Genome Database suite maintained by the USDA-ARS, and others both in the US and international locations are looking towards combining sequence computation with graphical displays of data return that can represent both intra- and inter-specific data. Inter-specific genome and germplasm queries will be greatly enhanced by access to common controlled vocabularies, and metabolic databases, both inter-plant and inter-all-genomes. At MaizeDB, we anticipate enhancing user-access to the data using menu-driven, user-defined table constructions. Instead of retrieving a list of loci with tassel phenotypes, one might specify a list of the loci, any PCR primers for nearby sites, gene products, GenBank accessions, the map bins. In the near future, we anticipate an onslaught of highly structured data from enhanced funding for crop plant genomes, both US and international.

REFERENCES:

1. Emerson, R. A. 1923 Letter. from files of E. H. Coe, University of Missouri-Columbia MO 65211.

2. Emerson, R.A., Beadle, G. W. and A. E. Fraser. 1935. A summary of linkage studies in maize.Cornell Univ. Agric. Exp. Stn. Memoir 180:1-83.

3. Letovsky, S. and M. Berlyn. 1994. Issues in the development of complex scientific databases. Proceedings of the 27th Annual Hawaii International Conference on System Sciences.

4. Byrne, P. F., M. Berlyn, E.H. Coe, G. Davis, M. Polacco, D. C. Hancock 1995. Reporting and accessing QTL information in USDA's Maize Genome Database. J. Quant. Trait Loci 1995:1-3

APPENDIX 1: Fragment of Genera specifications for the Short Locus Query Form

All fields listed, but only certain ones shown for query. 'Query select' option provides a menu to the user of the available options. The forms are 'freshened' overnight and new options will appear at that time.

```
LocusLite
*ID#: integer
            -key
            -noquery
            Column: ID# Locus ID#
*Name: char
            -lookup
            -noquery
*Symbol, Fullname, or Synonym:
            -listsinglecolumn
            *Synonym: char
                        Column: ID# Locus#Synonyms Synonym
                        -match
            *Per: Person
                        Column: ID# Locus#Synonyms Authority
                        -noquery
*Type: Term
            Column: ID# Locus Type
            -queryselect
*Fullname: char
            -noquery
*Species
            -queryselect
*Detected By:
            *Probe
                        Column: Locus Locus#DetectedBy Probe
            *Method: Term
                        Column: Locus Locus#DetectedBy Method
                        -queryselect
*Gene Products: setof Gene Product
            Column: Locus Locus#GeneProducts GeneProduct
*Linkage Group
            Column: LinkageGroup
            -queryselect
*Map Scores: setof Map Scores
            Column: ProbedSite MapScores ID#
            -noquery
*Recombination Data: setof Recombination Data
            Column: Locus RecombinationData#Loci RecombinationData
            -noquery
*Arm: Term
            -queryselect
...
```

Appendix 2: part of the Query Form generated from the above specification.

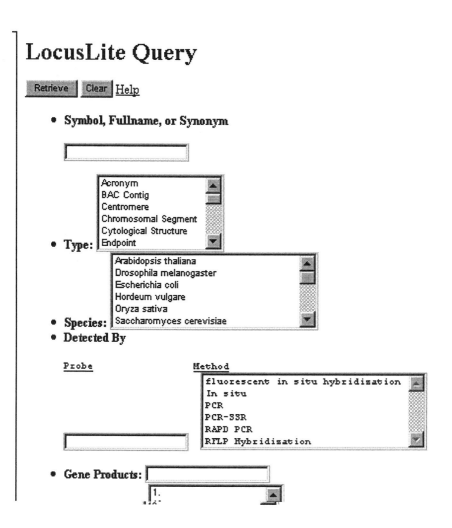

Appendix 3: Detail display fragment for the wx1 waxy 1 locus, also generated according to the specification in Appendix 1.

LocusLite wx1 waxy1

- **ID#:** 12768
- **Name:** wx1
- **Symbol, Fullname, or Synonym**

Synonym	Per
umc25	
umc25(wx)	
wx1	Canonical Name
Gss1	Plant-Wide Name
waxy1	Full Name
phi022	Senior, L
phi027	Senior, L
phi061	Senior, L
npi16-wx1	Wright, S
gsy200(wx)	INRA

- **Type:** Gene
- **Fullname:** waxy1
- **Species:** Zea mays ssp. mays
- **Detected By**

Probe	Method
p-pBF225	RFLP Hybridization
p-umc25	RFLP Hybridization
p-phi022	PCR-SSR
p-phi027	PCR-SSR
p-phi061	PCR-SSR
p-pBF224	RFLP Hybridization
p-p4Z-Wx	RFLP Hybridization

14 AGIS: Using the Agricultural Genome Information System

Stephen M. Beckstrom-Sternberg * and D.
Curtis Jamison**

*Department of Plant Biology, University of Maryland,
College Park, MD*

** Present Address: NIH Intramural Sequencing Center,
Advanced Technology, Center, National Institutes of
Health, Gaithersburg, MD 20877 (stevebs@nhgri.nih.gov)*

*** Present Address: National Human Genome Research
Institute, National Institutes of Health, Bethesda, MD.
20715 (cjamison@nhgri.nih.gov - corresponding author)*

Introduction

The Agricultural Genome Information System (AGIS – http://probe.nal.usda.gov/)
provides Internet access to genome information from agriculturally important
organisms. The server delivers information from thirty-six databases, encompassing
mostly crop and livestock animal species, including the databases for all of the major
food crop genome projects. Also included are a number of databases which have
related information, such as databases for several model organism genome projects
(including ACeDB for the nematode, *Caenorhabditis elegans*, and DictyDB for soil
amoebae, *Dictyostelium discoidium*), reference databases such as Mendel (plant gene
nomenclature), PhytochemDB (plant phytochemicals), EthnobotDB (plant uses), Fire
Ant, and OMIA (Online Mendelian Inheritance in Animals), as well as links to other
important resources like AGRICOLA. A complete list of databases is shown in
Table 1.

Table 1 : AGIS Databases

Plant Genome

> AAtDB--*Arabidopsis*
> Alfagenes--alfalfa (*Medicago sativa*)
> BeanGenes--*Phaseolus* and *Vigna*
> CabbagePatch--*Brassica*
> ChlamyDB--*Chlamydomonas reinhardtii*
> CoolGenes--cool season food legumes
> CottonDB--*Gossypium hirsutum*
> GrainGenes--wheat, barley, rye and relatives
> MaizeDB--maize
> MilletGenes--pearl millet
> RiceGenes--rice
> RoseDB--Rosaceae
> SolGenes--Solanaceae
> SorghumDB--*Sorghum bicolor*
> SoyBase--soybeans
> TreeGenes--forest trees
> Mendel--plant-wide gene names

Livestock Animal Genome

> BovGBASE--Bovine
> ChickGBASE--poultry
> PiGBASE--swine
> SheepBASE--sheep
> OMIA--Online Mendelian Inheritance in Animals

Other Organisms Genome

> ACeDB--*C. elegans*
> DictyDB--The soil amoebae *Dictyostelium discoideum*
> MycDB--*Mycobacteria*
> PathoGenes--fungal pathogens of small-grain cereals
> RiceBlastDB--the rice blast fungus *Magnaporthe grisea*

Plant Reference

> AGRICOLA--plant genetics subset
> CIMMYT--Wheat International Nursery Data
> Ecosys--plant ecological ranges
> EthnobotDB--worldwide plant uses
> FoodplantDB--Native American food plants
> MPNADB--medicinal plants of Native America
> PhytochemDB--plant chemicals
> PVP--Plant Variety Protection
> PVPSoy--Soybean Plant Variety Protection Data

Insect Reference

> Fire Ant--*Solenopsis*
> Face Fly--*Musca autumnalis*

Horn Fly--*Haematobia*
Screwworm--*Cochliomyia hominivorax*
Stable Fly--*Stomoxys calcitrans*

AGIS uses the World-Wide Web (WWW) technology to distribute information. Any person with a forms-capable browser can access the databases. The forms interface was kept as simple as possible, using only standard HTML commands, so as to maintain compatibility with as many WWW browsers as possible. All current versions of Netscape browsers (Navigator and Communicator) work, as does Internet Explorer and even old versions of Mosaic.

Implementation

ACEDB, the genome database developed for use with the *C. elegans* genome project [1], is used as the back-end. While other commercial database systems were initially considered, ACEDB is the *de facto* standard for genome projects. Thus, by using ACEDB, data compatibility with the AGIS collaborators was maintained.

Genome data delivery from AGIS has evolved, keeping pace with new computer technologies. Originally, information was distributed on the Internet by a Gopher server [2], and by CD-ROM for researchers who had no Internet access. Both methods were discontinued as AGIS was migrated to the World-Wide Web by the use of a highly modified ACEDB program [3], which generated static HTML pages for text displays and GIF images for graphical maps. The current, more interactive WWW interface is provided by a package called webace [4], which utilizes forms-based HTML pages to provide more access to ACEDB functions, such as the Query by Example and Table-maker facilities, as well as interactive GIF images.

The architecture of AGIS is shown in Figure 1. The databases are incorporated into the aceserver layer, which runs as an inetd daemon using RPC calls. The aceserver is a modified form of ACEDB, and includes the giface program, which turns ACEDB graphics into GIF images suitable for use by webace.

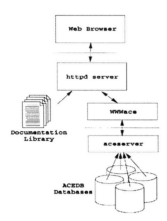

Figure 1: AGIS Architecture

The webace program is a PERL script that runs as a CGI program, utilizing the CGI.pm PERL module. Webace translates user queries into the ACEDB query language, and then converts ACEDB objects into HTML documents for display. HTML links are created by a set of simple markup rules [5], and can insert URL anchors which point to other ACEDB objects, external database queries, or even external analysis programs like the NCSA Biology Workbench [6].

Guided Tour of the AGIS Databases

From the main AGIS, menu (Figure 2), the user can access the list of available databases by following the "Databases" link. Databases are grouped according to type (plant, livestock, model organism, and reference). Following each of the database listing is a set of links (browse, query, about), shown in Figure 3, which allow the user to jump to the particular HTML forms set to access data from that particular database.

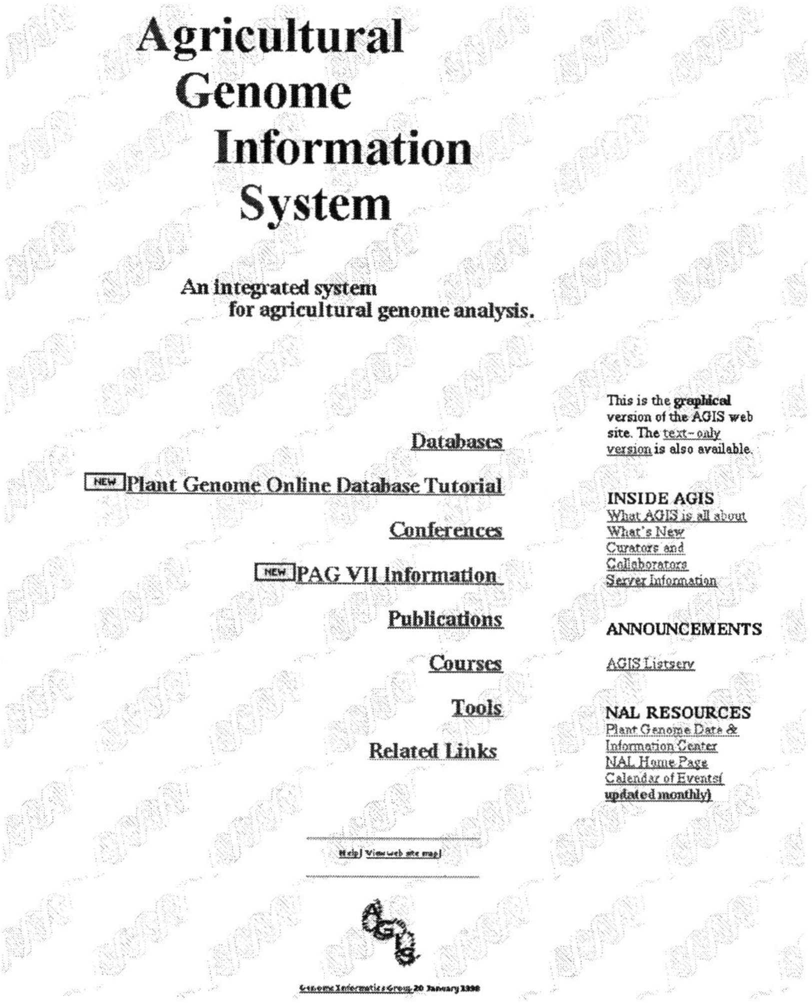

Agricultural Genome Information System

An integrated system for agricultural genome analysis.

This is the **graphical** version of the AGIS web site. The text-only version is also available.

Databases

NEW Plant Genome Online Database Tutorial

Conferences

NEW PAG VII Information

Publications

Courses

Tools

Related Links

INSIDE AGIS
What AGIS is all about
What's New
Curators and
Collaborators
Server Information

ANNOUNCEMENTS

AGIS Listserv

NAL RESOURCES
Plant Genome Data &
Information Center
NAL Home Page
Calendar of Events(
updated monthly)

Help| View web site map|

Genome Informatics Group 20 January 1998

Figure 2 : Main AGIS Menu

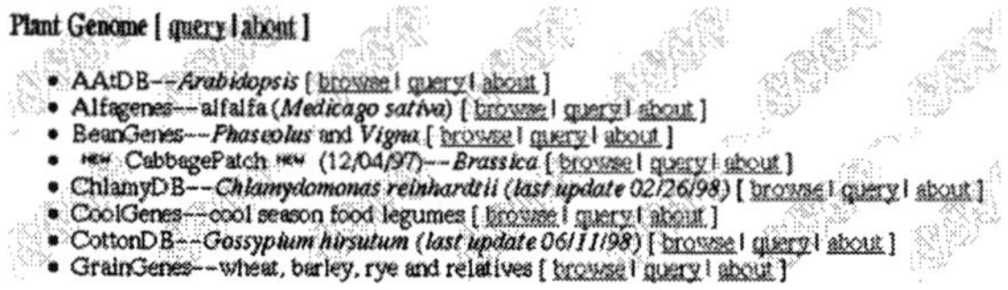

Plant Genome [query | about]

  • AAtDB--*Arabidopsis* [browse | query | about]
  • Alfagenes--alfalfa (*Medicago sativa*) [browse | query | about]
  • BeanGenes--*Phaseolus* and *Vigna* [browse | query | about]
  • NEW CabbagePatch NEW (12/04/97)--*Brassica* [browse | query | about]
  • ChlamyDB--*Chlamydomonas reinhardtii* (*last update 02/26/98*) [browse | query | about]
  • CoolGenes--cool season food legumes [browse | query | about]
  • CottonDB--*Gossypium hirsutum* (*last update 06/11/98*) [browse | query | about]
  • GrainGenes--wheat, barley, rye and relatives [browse | query | about]

Figure 3 : An Example of AGIS Database Links

There are two basic access methods for the plant genome data in AGIS: 1.) Browsing interface; 2.) and Query interface. Each method has advantages and disadvantages, and the interface of choice will be strongly dependent upon both the data desired and the familiarity of the user with ACEDB. Database access is restricted to single databases at present -- no concurrent or cross-database querying is supported. The following sections present the three interfaces arranged by complexity and power. Finally, a section on the hypertext ACEDB objects is presented.

Browse Mode

For novice users, the browse mode is certainly the simplest interface. Browse is a point and click interface which allows the user to wander through the data in a completely hyper-linked mode. A problem with this approach is the complexity and amount of the data can be overwhelming. Still, using the browse mode allows the user to avoid the complexities of the ACEDB query language.

Selecting the browse mode presents the user with a form listing the available classes in the database. The available classes choice page for the RiceGenes database is shown in Figure 4. Selection of a class (in this case Locus) presents the user with a list of all objects available for that class. When there are too many objects to be comfortably listed, the list is collapsed into a set of sublists. Selecting a sublist (Figure 4) brings up a shortened list of all objects in that particular range. The user always has the option to import the entire list.

Selection of an object from the final list retrieves a hypertext version of the object, as described below.

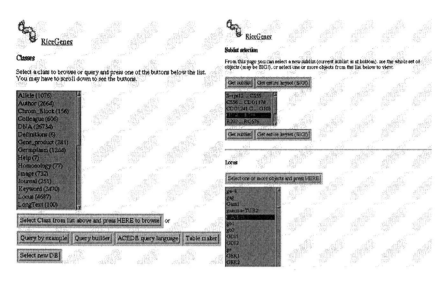

Figure 4 : Available Classes and Sublist Selection Pages for RiceGenes

Query Mode

Selecting query mode brings the user to a form allowing the databases to be searched/queried by six different methods, including fuzzy search, WAIS, Query by Example, Query Builder, Table-maker, and ACEDB Query Language in which an ACEDB query can be input directly.

ACEDB query. This mode is extremely fast and efficient for data retrieval, but it does require the user to be familiar with the ACEDB query language as well as the data structure of the particular database. While these two topics are outside the scope of the current paper, the reader is referred to several excellent treatises on the ACEDB query language [7,8], available at the AGIS site.

170

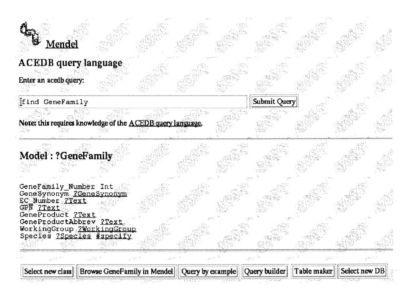

Mendel

ACEDB query language

Enter an acedb query:

| find GeneFamily | | Submit Query |

Note: this requires knowledge of the ACEDB query language.

Model : ?GeneFamily

```
GeneFamily_Number Int
GeneSynonym ?GeneSynonym
EC_Number ?Text
GPN ?Text
GeneProduct ?Text
GeneProductAbbrev ?Text
WorkingGroup ?WorkingGroup
Species ?Species #specify
```

| Select new class | Browse GeneFamily in Mendel | Query by example | Query builder | Table maker | Select new DB |

Figure 5 : ACEDB Query Interface

Figure 5 shows the basic ACEDB query interface. The query is typed into the form and submitted, and a hypertext ACEDB object is returned.

Fuzzy (AGREP) and WAIS query. The fuzzy and WAIS modes present the simplest query interface. One or more words can be entered as the search string, and wildcards are accepted. A search returns a hypertext list of database objects.

Query by Example. This mode allows the user to query one class of a database by typing search strings into one or more field categories on a form. The query brings back a list of matching objects. Figure 6 shows the Query by Example interface to the GeneFamily class of the Mendel database.

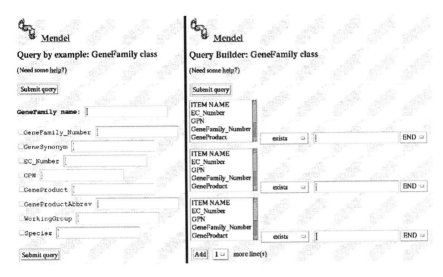

Figure 6 : Query by Example and Query Builder Interfaces

Query Builder. The Query Builder interface allows much more complex queries than Query by Example, permitting the user to string together any number of "and", "or", and "xor" queries together for a particular class from a database. Figure 6 illustrates the Query Builder interface for the GeneFamily class of the Mendel database.

Table-maker Mode. The most powerful method of accessing data from ACEDB data is the Table-maker program. Table-maker allows the user to create a relational database style table of ACEDB objects. Like the query interfaces, Table-maker requires an understanding of the structure of the database. However, the challenge of acquiring this knowledge is more than made up for in improved information retrieval.

The AGIS interface to the ACEDB Table-maker utility is shown in Figure 7. Table-maker works by making an initial query, then applying modifiers and performing "follow" operations, which can be thought of as automated linking operations between objects.

A nice tutorial for Table-maker can be found at the AGIS web site [9].

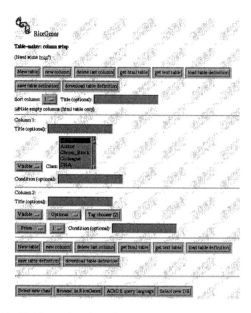

Figure 7 : Table-maker Interface

Hypertext ACEDB objects

Regardless of the mechanism used to initially query the database, an ACEDB object is the ultimate goal of interacting with the databases. A typical ACEDB object is shown in Figure 8. The object looks very similar to a native ACEDB object, with links to other objects represented as hyperlinks. Following such a link will replace the current object by following the link.

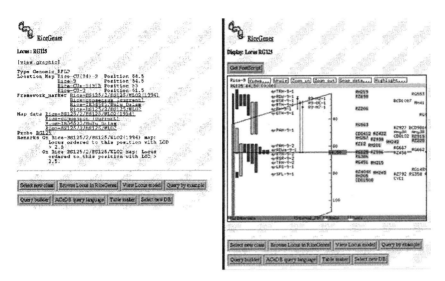

Figure 8 : Hypertext ACEDB Object and ISMAP from RiceGenes

One salient difference between native ACEDB objects and those presented by the AGIS system is that while ACEDB objects launch multiple viewing types (e.g., maps, text, images), the default type of any object returned by the AGIS system is text. At the top of each object is the "View as Graphic" link, which invokes the GIFace server and returns a clickable ISMAP graphic (Figure 8). While mildly annoying to those looking specifically for graphical views, the text orientation of the AGIS server was a conscious decision designed to conserve bandwidth and avoid burdening users who have slower network connections with unwanted graphics.

The Future of AGIS

New technologies bring new opportunities to improve service, and the providers of AGIS have always striven to make use of as much state-of-the-art technology as possible. One such new technology is the Java programming language which allows interactive viewers and programs to be downloaded across a WWW link. JAVA holds promise to improve the AGIS interface, allowing it to be much more interactive and responsive than permitted by standard HTML forms.

Jade is the Java version of ACEDB [10]. A new experimental AGIS server has been created, with small modifications to the basic Jade structure to give it a familiar look and feel. While the basic interactions are similar to the WWW version, the Java map viewers are a great improvement. Additionally, the Java version for the first time holds forth the promise of concurrent database queries and cross-species comparisons performed directly from the AGIS servers.

174

Acknowledgments

AGIS is the product of many talented people, including J. Barnett, D. Bigwood, S. Cartinhour, G. Juvik, J. Krainak, T. Le, J. Martin, M. Shives, M. Sikes, and D. Vo. AGIS is a cooperative effort between the USDA Plant Genome Initiative and the University of Maryland, and is housed by the USDA National Agriculture Library. This work was supported by the USDA-NGRP National Plant Genome Research Program.

References

1.   Durbin, R. and J. Thierry-Mieg (1991-). *A C. elegans Database.* Documentation, code and data available from anonymous FTP servers at lirmm.lirmm.fr, cele.mrc-lmb.cam.ac.uk and ncbi.nlm.nih.gov.

2.   Alberti, B., F. Anklesaria, P. Lindner, M. McCahill, and D. Torrey (1991). *The Internet Gopher Protocol*, available at, gopher://boombox.micro.umn.edu:70/00/gopher/gopher_protocol/protocol

3.   Decuoux, G. (1995). A webserver for ACEDB, available by ftp from moulon.inra.fr in /pub/www acedb .

4.   Barnett, J.D. and D.W. Bigwood (1996). *WWW interface to ACEDB*, available at, ftp://probe.nal.usda.gov/pub/tools/webace.tar.gz
5.   Barnett, J.D., D.W. Bigwood, and S. Cartinhour. (1995*). A World-Wide Web Server for ACEDB based on Tace*, available at http://probe.nalusda.gov:8000/acedocs/ace95/nalwww.html
6.   Unwin, R., J. Fenton, M. Whitsitt, D. C. Jamison, M. Stupar, E. Jakobsson and S. Subramaniam "The Biology Workbench: A WWW-based Virtual Computing and Analysis Environment for Macromolecular Structures and Sequences." (this volume).
7.   Matthews, D, and S. Lewis. 1995. *Searching an ACeDB database,* available at http://greengenes.cit.cornell.edu/acedoc/query.syn.html
8.   Barnett, J. 1995. *ACEDB Query Language Examples*, available at http://probe.nalusda.gov:8000/acedocs/acequery_examples.html .
9.   Barnett, J. 1995. *Introduction to Table-maker on the WWW*, available at http://probe.nal.usda.gov:8000/acedocs/cimmyt_workshop/table-maker.html .
10.   Stein, L., J. Thierry-Mieg, S. Cartinhour. Jade paper (this volume).

15 CGSC: THE *E.COLI* GENETIC STOCK CENTER DATABASE

Mary
K.B. Berlyn

E. coli Genetic Stock Center, Dept. of Biology, 355 OML, 165 Prospect St.
New Haven, CT 06511

Introduction

The *E. coli* Genetic Stock Center was founded 25 years ago for the purpose of collecting and distributing useful genetic derivatives of *E. coli* K-12 made by and used for teaching and research by geneticists, molecular biologists, biochemists, and others in academic, medical, government and industry laboratories. Historically, the Stock Center had never aspired to be a broad informatics resource, but in fact a full and accurate description of experimental strains and information needed for effective use of the strains involves detailed descriptions of the genotype, and in many cases phenotypic characteristics resulting from the genotype, of mutations and their properties, of gene function (the RNA and polypeptide gene products and also phenotypically defined functions), of gene map locations, and also of supporting information on pedigrees, references, people contributing and using strains, and feedback on strain phenotype. Over 7500 strains have been officially accessioned into the stock center over the years and about 4000 of these are part of the current working collection. Many of the strains contain 12-25 different mutations. As a result, the database includes over 3700 distinct mutations, over 2200 genes, with information on their organization into several hundred operons, about 600 sites of insertion for transposon mutations, prophage integration, etc., replication origins and termini, identification of 1400 RNA or protein products encoded by the genes, and 8000 references documenting genes, gene products, or strains. See http://cgsc.biology.yale.edu.

Development and Features of the Database

For 20 years, this information was held in very organized format by a single curator, Dr. Barbara Bachmann, but in the form of cross-referencing filecard catalogs, notebooks containing pedigree diagrams, genotype descriptions, gene and gene function tables and allele information, and also, informal notes and human memory. Because of its role in tracking genes and alleles, the stock center has also taken on the task of registering alleles and of publishing the linkage map for *E. coli* K-12 since 1976 (e.g., 1-4). Converting this information to electronic form was a task begun in 1989 as a two-phase development that was functional, in terms of software and essential data entry, in early 1990. A major imperative for this was the need to ensure the continuity of the stock center into the future, and the crucial need to modernize the records as part of this process had been recognized by program and division officers at the supporting agency, the National Science Foundation, for some time. The structured and generally consistent nature of the record-keeping that had evolved during those 20 years, the clear mission of the stock center, the observable patterns of usage of the various types of data, and the absence of a pressing deadline for completion facilitated a user's needs and dataflow analysis that led to conceptual and data models. We wanted the robustness of a commercial relational database management system and eventually chose Sybase from among those available, while keeping the model itself 'object-oriented'. Some schema modifications occurred during the implementation phases (with Stan Letovsky the sole software developer for a rapid development and testing process), but the resultant database bore a striking, perhaps surprising, resemblance to the early plans.

Several aspects of the conceptual data model had either not been included in models of other databases or were distinctly different from other treatments. Any segment of the canonical (wildtype) chromosome was modeled as a "Site" (alias locus). This includes genes, control regions, intergenic and intragenic regions, groups of genes, including operons, segments of the chromosome that were deleted, inserted, or inverted in structural mutations. Every site has a left endpoint and a right endpoint. Thus overlaps between the end of one gene and the beginning of another or inclusion of a regulatory region within an adjacent gene could be described and detected in searches. The coordinates for these points can have multiple values, reflecting different map versions, with the current version, of course, setting coordinates according to completed nucleotide sequence for *E. coli* K-12 (5). Sites can have "subsites"; e.g., all the genes carried on a deleted or inverted segment are subsites of the segment, and genes within an operon are subsites of the operon, with each subsite also being represented as an independent site. Since there was a single isolate of K-12 that is considered to be THE "wildtype", the structure and sequence of this wildtype chromosome can be used to define the standard chromosome (genome), and deviations from this structure can be described as mutations of the wildtype. Only mutations need be described in presenting the genotype of a strain; this is the convention followed by geneticists since the earliest days of the field. It is important in the database structure, since properties that belong to the gene itself can be described once for that gene and need not be repeated for all the various mutations of that gene. In the database, every mutation is linked to a Site in the relationship Mutation OF: Site, and for intragenic changes, the mutations are alleles OF a gene; for structural mutations, they are, for example, deletion-type mutations OF a "chromosomal region"-type site.

Since we did not wish to be redundant in providing information already available in existing databases, we chose to make links between the CGSC Sites and sequences in GenBank, between CGSC References and Medline records that include abstracts, and between Gene Products and Enzyme Commission and SwissProt databases. This was a pre-WWW decision and fortunately, the development of first Mosaic and then Netscape and other browsers and the resultant expansion in Web use and convenience validated this as a workable strategy.

Re-usable parts of the project

In a sense parts of the data model can be considered "re-usable" parts of the project, since discussions of this model were influential in adoption of similar, but modified models for aspects of both the *Maize Genome Database* (6) and its component Maize Stock Center information and the *Arabidopsis Information Management System* (7) at the Arabidopsis Biological Resources Center. Work with the maize database group at the University of Missouri gave impetus to the development of Stan Letovsky's *Genera* software (8) which has allowed development of a Web interface and easy modification methods for the database. *Genera* also has been used for creation of other databases with Sybase backends and Web frontends (e.g., *Field Guide to Puerto Rico* by Todd Forrest, ref. 9), so this is a re-usable spinoff from the initial database projects.

Query Examples

Probably the most used, most valuable, and most distinctive query used by public users on the web and by stock center personnel either on the web or on the aptforms is looking for all strains that have specific combinations in their genotype. For example, a strain that carries an F-factor, and is restriction-negative (*hsdR⁻* and RecA⁻, [or carries specific auxotrophic markers] can be found by selecting F⁺, F′, and Hfr in the Sex field of the **Strain Query** form and entering the gene symbol or mnemonic for the type of allele being sought in the Mutations field. This is shown in Figure 1. The upper left section illustrates the **Strain Query** form and the appropriate entries, to the right, the list of strains retrieved from that query and the below, the description on the **Strain** form of the strain selected from the query result.

Another kind of query is illustrated, accessible from the Aptforms interface, is shown in Fig. 2 It asks for a strain that carries a *lacZ* amber suppressible mutation and is *lacI⁻* and also for a strain that is isogenic to this strain, except for the *lacZ* marker. This is a complex, but frequently needed query for experimentalists, and it is facilitated by furnishing a field entitled "Isogenic with respect to mutation:" Fig. 2 shows the entry for the query and the list of strains that have the required markers and also have an isogenic partner in the collection. After browsing the genotypes of the retrieved strains, the user may select one or both; the description of the selected strain is shown and the arrow indicates the "Isos!" command which retrieves its isogenic partner as well. The "Report" function formats the retrieved strain descriptions in a more formal way, shown at the bottom of the figure.

FIGURE 1. A Query for an *hsdR⁻ recA⁻* strain also having an F-factor

---

Strain Query
	Retrieve	**Clear**		**Help**

CGSC#: _____
Strain Designation
| **Designation** | **Source Person** |

_____   _____

Sex(Hfr,F+,F-, or F'): Hfr F+ F'
PO#: _____   **PO_Map_Position:** __ +/-__
PO_is_Clockwise(1=CW,0-CCW): _____
Episome: _____
Mutations:
> hsdR
> recA

No. of Muts Carried: ____
Special Uses: [select list offered]
Comments: _____
External DB Keys: _____
References: _____

⇒   2 items retrieved
DH20   **DH21**
    [select DH20]

⇓

Strain: DH20
CGSC#: 6818
Designation:	**Source Person**	**Choice**
DH20	Hanahan, D.	1

Episome: F128-18
Mutations	**Certainty**

> *glnV44(AS)*
> *lambda⁻*
> *rfbD1*
> *gyrA96(NalR)*
> *recA1*
> *relA1*
> *endA1*
> *spoT1*
> *thi-1*
> *hsdR17*

No. of Muts Carried: 10
Comments: This F-prime comes from the strain X91, which carries a derivative of F128-10 that was designated F128-17. F128-17 & F128-18 are probably the same F', extending from (*proB-lac*) and carrying *lacIp-4000(lacIQ)*. NalR=nalidixic acid resistance.
Reference: Hanahan, D. 1983. Studies on transformation of *Escherichia coli*   with plasmids. J.Mol.Biol. 166:557-580.
    ID#: 7408

In this query "Amber" is a mutation property, and certain properties, such as constitutivity or nonsense suppressibility (e.g., Amber) are such distinctive properties that traditionally a symbol is parenthetically appended to the mutation designation. This convention fortuitously provides a shortcut for this query. It is not necessary to "subquery" on the Mutation form to specify the property, but only to enter it as part of the name, "*lacIZ*wildcard](Am)". The retrieval on these forms also illustrates another feature of the database. Some comments (shown as Type Inherited on the form) are Mutation Comments, inherited by all strains carrying that mutation. Also, the definition of the Mutation property (Am) is inherited by the Strain form.

Other common queries are for strains that carry a specific mutation with a transposon or other selectable marker nearby; a strain that carries a mutation and is an Hfr that transfers that mutation early in a mating; a combination of mutations in two or more specific genes; or a strain that lacks a specific enzyme. These are easily carried out by users familiar with the genetics of E. coli.

A current version of the traditional linkage map ("stick-and-tick" representation) can also be accessed from the web, as drawings of 2-minute segments from the web site, or as an ordered list of genes and coordinates drawn directly from the database.

Quality Control and Curation

Because we were confronted with so much data entry from paper files, we initially emphasized triggers and tools that attempted to facilitate and ensure data integrity for entries and updates, and these have served us well in maintaining quality control and in the curation process.

We attempted to circumscribe the content of this database, so that the part-time attention of one or two people could accomplish curation of the "primary data", and access to important related data would be provided by links to other data

FIGURE 2. A Query for a *lacZ*-(amber) *lacI⁻* strain, with no amber suppressor, and an isogenic *lacZ⁺* strain

---

STRAIN
Select! **Sub!** **Clear!** **Go!** **Clone!** **Report** **Pedig!** **Isos!**
[?] ID#:___ CGSC#___ Designation Choice Source
Sex: ____ _____ __ _____
Episome: _____ Via Use
Mutations -------------- ----------------
[lacI%] Isogenic w.r.t. mut: [lacZ%(Am)]
[(AS)] Comment: _____
References: Author____Year___Journal_____Vol. ___Page __Priority

---

⇓

STRAIN
  Select! **Sub!** **Clear!** **Go!** **Clone!** **Report** **Pedig!** **Isos!**

[!] **ID#:** 7668 **CGSC#** 4977 **Designation Choice Source**
3.300 U281 1 Paris
Sex: Hfr CA86 2 S.Brenner
PO: valS < attP4 **Via Use**
Episome: _____ __ _____
Mutations
Mutation	**Type**	**Comment**
lacZ281(Am)		
lacI22		
LAM-	Inherit	lacI22 was formerly called lac$_{i\,3}$ by Jacob; it originated
relA1		in strain 3.300 Hfr.
spoT1	Inherit	lacZ281 was formerly called lacZ$_{U281}$.
thi-1	Inherit	Am = amber mutation

References:
Author	**Year**	**Journal**	**Vol.**	**Page**	**Priority**
Pardee et al.	1959	J.Mol.Biol.	1	165	1

⇒ **Isos!** and **Report!** ⇒

CGSC Strain # 4977 **Strain designation:** 3.300 U178
Other Designation: CA86 S.Brenner
Sex: Hfr **Point of Origin** 1 of Hfr 3000 *valS--<--attP4*
Chromosomal Markers:
  lacZ281(Am), lacI22, l⁻, relA1, spoT1, thi-1
Comments: *lacI22* was formerly called lac$_{i3}$ by Jacob; it originated in strain 3.300 Hfr.
lacZ281 was formerly called lacZ$_{U281}$. Am=amber(UAG) mutation
Reference: Pardee et al. 1959J.Mol.Biol. 1:165

CGSC Strain #808 **Strain Designation:** 3.300
Sex: Hfr **Point of Origin** 1 of Hfr 3000 valS--< attP4
Chromosomal Markers:
  lacI22, l⁻, relA1, spoT1, thi-1
Comment: *lacI22* was formerly called lac$_{i3}$ by Jacob; it originated in strain 3.300 Hfr.

sources --- sequences (GenBank), bibliography abstracts (Medline), and more detailed enzyme information (Enzyme Commission database and SwissProt). This choice reflects the fact that the stock center has always operated with a small staff, usually director, two laboratory research assistants, some part-time help with data entry and editing, and with the advent of the database, a part-time systems administrator, and the database is only an ancillary part of the stock center's functions. The body of data, however, expands and needs modification at a rate that exceeds our modest plan. We need to adjust our curation model to accommodate this.

Lessons learned and improvements obvious in hindsight

One of the positive lessons was that one biologist beginning at ground level in database experience and one computer scientist with interest and knowledge in biology can work together to accomplish a lot of database functionality in a short time.

On the negative side, I would, in hindsight, pay more attention to external use of the database from the outset. The initial development definitely focussed on in-house use because that was the most critical need and because external use was not favored by the then-director of the stock center. Subsequently, we have found external access by users to be extremely helpful and efficient both for users and for stock center staff. The ability to provide a Web interface to the database (10), thanks to the Genera software (8), has done a lot to increase satisfaction of outside users. Yet there are many useful features that we use daily with the in-house aptforms that are not yet available to web users. There are ways that we can accommodate some of these features, even within html limitations, but they have not all been attended to. Among the things I would do differently would be to ensure that external users had access to these more powerful querying capabilities from the outset (although certain parts of the database, such as request-forms and strain-maintenance records, would still be excluded from the public part of the database, since they seem irrelevant to external users' information needs). Also, of course, we would develop more extensive user documentation.

External use is also related to curation capabilities. Unexpectedly, being on the web has given us a volunteer corps of proofreaders. Many users are experts in areas of biochemistry or genetics and send us corrections on very specific aspects of the data. It has been suggested by some outside users that there be public access to:
(a) a write-in adjunct to the database to allow researchers to add expert knowledge, and corrections, that will become appended to the record and available to anyone examining the record.
(b) allowing some SQL-query capability (non-forms).

These are suggestions that deserve serious consideration. However, the rate of daily use, and the fact that most scientists requesting strains indicate that they have successfully examined the database on the web prior to asking for strains or further information, very often finding the specific strain they need, has convinced us of the usefulness of the database in providing strains to the research community.

Acknowledgments

In the planning process, I was extremely fortunate to have the help and encouragement of program and division officers then at the National Science Foundation, including Drs. Robert J. Robbins, James Edwards, and John Wooley, Drs. Gerald Selzer, David Kingsbury, and others. A CGSC database advisory group also provided valuable help in the early planning and modelling phase and this group included Drs. Jim Ostell of NCBI, Tom Marr then at LANL, Ken Sanderson of U.of Calgary, Brooks Low of Yale, and R. Robbins then of NSF, upon occasion augmented by other software experts. I am also fortunate that the software developer for the project was Stan Letovsky, and Peter Kalamarides is systems administrator. The richness and availability of a very large segment of the data are the result of the dedication and expertise of Barbara Bachmann, during her long tenure at the Stock Center. I'm grateful for the insights and suggestions of research assistants Linda Mattice and Narinder Whitehead, who have become major users of the database in their daily activities, and for those of scientists whose more occasional use also led them to offer criticism, suggestions, and requests for improvements. This work was supported by the National Science Foundation.

References

1. Bachmann, B.J., Low, K.B., and Taylor, A.L. Recalibrated linkage map of *Escherichia coli* K-12. Bacteriol. Rev. 40:116-167 (1987)
2. Bachmann, B.J. Linkage map of *Escherichia coli* K-12, edition 8. Microbiol. Rev. 47:180-230 (1990)
3. Berlyn, M.K.B., Low, K.B., and Rudd, K.E. . Linkage map of *Escherichia coli* K-12, edition 9, p. 1715-1902. In F.C.Neidhardt, et al. editors, *Escherichia coli* and *Salmonella* : cellular and molecular biology, 2nd ed. ASM Press, Washington DC (1996)
4. Berlyn, M.K.B. Linkage map of *Escherichia coli* K-12, edition 10. In Press. Microbiology and Molecular Biology Reviews. (Sept. 1998)
5. Blattner, F. et al., The complete genome sequence of *Escherichia coli* K-12. Science 277: 1453-1474 (1997)
6. http://teosinte.agron.missouri.edu/top.html
7. http://aims.cps.msu.edu/aims/
8. Letovsky, S. Genera: http://cgsc.biology.yale.edu/genera
9. Forrest, T. http://cgsc.biology.yale.edu/newfield.html
10. Berlyn, M. http://cgsc.biology.yale.edu
11. Berlyn, M. Accessing the *E. coli* Genetic Stock Center Database. p. 2489-2495. In F.C.Neidhardt, et al. editors, *Escherichia coli* and *Salmonella* : cellular and molecular biology, 2nd ed. ASM Press, Washington DC (1996)

SYSTEMS

16 OPM: OBJECT-PROTOCOL MODEL DATA MANAGEMENT TOOLS `97

Victor M. Markowitz, I-Min A. Chen, Anthony S. Kosky, and Ernest Szeto

Data Management Research and Development Group
Lawrence Berkeley National Laboratory, Berkeley, CA 94720[*]

Introduction

The development of the Object-Protocol Model (OPM) and OPM data management tools started in 1992. The main motivation for developing OPM at the time was the need to provide data management support for large scale DNA sequencing laboratories. We designed the *protocol class* as a construct for modeling sequencing, as well as other scientific, experiments. Since object-based data models were well suited for modeling the complex data underlying the scientific and traditional database applications targeted by OPM, we decided to incorporate the protocol class construct into the framework of an object data model [2].

The development of the OPM data management tools can be split into three stages. Initially, the OPM data management tools aimed only at providing *wrapper* facilities for developing and querying individual databases implemented with commercial relational database management systems (DBMSs). The OPM Database Development and Database Query tools provided such facilities, first for Sybase and later Oracle, both widely used for implementing large production biological databases. These tools were employed for developing and maintaining several biological databases, such as version 6 of the Genome Data Base (GDB)[1] at Johns

---

[*] Current affiliation: BIOINFORMATICS SYSTEMS, GENE LOGIC, INC., 2001 Center Str., Suite 600, Berkeley, CA 94704.
Email: {vmmarkowitz, ichen, anthony, szeto} @ genelogic.com
[1] http://gdbwww.gdb.org/

Hopkins School of Medicine in Baltimore, and the Primary Database of the German Human Genome Resource Center (RZPD)[2] in Berlin, Germany.

Next, the OPM Retrofitting tools were developed in order to provide support for constructing OPM views for databases that were not originally developed using the OPM tools, and for providing the infrastructure required for using the OPM Database Query tools to access these databases. We developed retrofitting tools for relational DBMSs and applied them to databases such as the Genome Sequence Database (GSDB)[3] at the National Center for Genome Resources (NCGR). More recently, we have developed retrofitting tools for structured flat files such as GenBank.[4]

Finally, we took advantage of the ability to build uniform OPM views on top of diverse databases by developing the OPM Multidatabase Tools for querying and exploring multiple heterogeneous databases via native OPM schemas or retrofitted OPM views. These tools have been applied to the construction of a Molecular Biology Database Federation that includes GDB, GSDB, and GenBank.

In the remainder of this paper we will briefly overview the OPM data management tools and will discuss the experience gained in the past five years of developing and applying these tools to scientific database applications. Details of these tools and their underlying methodologies can be found in the OPM papers listed as references; most of these papers are available on the Web at http://gizmo.lbl.gov/opm.html.

The OPM project is currently at a crossroads. The OPM tools were developed by members of the Data Management Research and Development Group at Lawrence Berkeley National Laboratory, with funding from the Office of Biological and Environmental Research and the Mathematical, Information, and Computational Sciences Division of the US Department of Energy. In September 1997 the developers of the OPM tools joined Gene Logic Inc., forming its Bioinformatics Systems division, where the next generation of OPM data management tools will be developed. The future versions of the OPM tools will include enhancements of existing facilities, such as more powerful Web-based interfaces, as well as new facilities, such as mechanisms for integrating data management and analytical tools, and providing support for database evolution.

The Object-Protocol Model

The Object-Protocol Model (OPM) is the result of incorporating constructs for modeling scientific experiments (protocols) into an object data model. We will briefly review the main features of OPM below; details can be found in [1].

---

[2] http://www.rzpd.de/
[3] http://www.ncgr.org/gsdb/
[4] http://www.ncbi.nlm.nih.gov/

OPM is a data model whose object part is closely related to the ODMG standard for object-oriented data models [10]. Objects in OPM are uniquely identified by object identifiers, are qualified by attributes, and are classified into classes. Classes can be organized in subclass-superclass hierarchies, where multiple inheritance in such hierarchies is supported.

Attributes can be simple or consist of a tuple of simple attributes. An attribute can have a single value, a set of values, or a list of values. If the value class (or domain) of an attribute is a system-provided data type or a controlled-value class of enumerated values or ranges, then the attribute is said to be primitive. If an attribute takes values from an object class or a union of object classes, then it is said to be abstract.

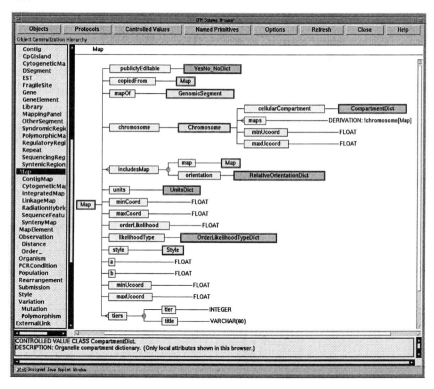

Figure 1: Part of the OPM Schema for GDB

Part of an OPM schema for GDB 6, viewed using the Java-based OPM Schema Browser, is shown in Figure 1, where the class Map is shown together with its attributes. For example, copiedFrom, mapOf and chromosome are abstract attributes with value classes Map, GenomicSegment and Chromosome respectively, while minCoord and maxCoord are primitive attributes, and includesMap is a tuple attribute with components map and orientation.

OPM supports the specification of derived attributes using derivation rules involving arithmetic expressions, aggregate functions (min, max, sum, avg, count) and attribute composition. In Figure 1, for example, attribute maps of class Chromosome is a derived attribute defined as the inverse of attribute chromosome of class Map. OPM also supports derived subclasses and derived superclasses. A derived subclass is defined as a subclass of one or more derived or non-derived object classes with an optional derivation condition. A derived superclass is defined as a union of two or more object classes.

In addition to object classes, OPM supports a protocol class construct for modeling scientific experiments. Similar to object classes, protocol classes have class names, optional class descriptions, identifiers, and are associated with attributes. Protocol modeling is characterized by the recursive specification (expansion) of generic protocols in terms of alternative subprotocols, sequences of subprotocols, and optional subprotocols. In addition to regular attributes, a protocol class can be associated with special input and output attributes that represent input and output data regarding the experiment modeled by the protocol class, and express input-output connections with related protocol classes.

Documentation in the form of descriptions, examples, and application-specific properties can be associated with OPM schemas as well as schema components such as classes and attributes. Further, classes can be organized into clusters, and clusters can be nested.

The OPM Query Language (OPM-QL) [3] is an object-oriented query language similar to OQL, the ODMG standard for object-oriented query languages [10]. An OPM query consists of a SELECT statement, specifying the values to be retrieved for instantiations of variables satisfying the query condition; a FROM statement, specifying the variables that occur in a query and the classes or attribute values which they range over; and an optional WHERE statement specifying conditions on instantiations, where conditions consist of and/or compositions of simple atomic conditions. A query may also involve local, inherited and derived attributes and path expressions starting with these attributes.

The Object-Protocol Model Data Management Tools

The OPM data management tools provide facilities for developing and accessing databases defined using OPM, for constructing OPM views of existing relational databases and structured files, for representing database schemas using alternative data-models, for publishing schemas in various formats, and for querying databases through uniform OPM views. The OPM multidatabase tools provide facilities for exploring multiple heterogeneous databases that have either native OPM schemas or retrofitted OPM views. We will briefly describe each of the OPM data management tools below.

The OPM Database Development Tools

OPM schemas can be specified using either a regular text editor or using the graphical OPM Schema Editor. The OPM Schema Editor is implemented in Java and provides a graphical interface implemented using the Java Abstract Windowing Toolkit (AWT). The tool allows object and protocol structures to be specified incrementally by defining new OPM classes, modifying existing OPM classes and defining attributes of classes. OPM schemas can be also examined graphically on the Web using the OPM Schema Browser. The OPM Schema Editor and the OPM Schema Browser provide facilities for generating Postscript diagram, LaTeX document and HTML file representations of OPM schemas.

OPM schemas are maintained as ASCII files that can be passed to the OPM Schema Translators described below, in order to generate the corresponding DBMS-specific database definition and constraints.

Individual OPM schemas or several related schemas can be documented in a Database Directory and Schema Library (DD&SL). A DD&SL contains information on individual databases, such as database names, underlying DBMS, access information, and the database schemas represented in OPM and other alternative notations, such as the Extended Entity-Relationship (EER) model, the relational model, and the ASN.1 data exchange notation. In addition the DD&SL contains information about the relationships between databases represented in the DD&SL, or *inter-database links*, which can be used in exploring across databases and formulating multi-database queries. The DD&SL is used by the Multidatabase Query Tools described later, in order to provide the information necessary formulating multidatabase queries. The information in a DD&SL may also be automatically converted to a hierarchy of HTML pages, so that the DD&SL may also be browsed using a Web browser, such as Netscape or Explorer (see [6] for details).

The OPM Schema Translator translates OPM schemas into relational database definitions and database procedures implementing the OPM retrieval and update methods [2]. Informally, the translation of an OPM schema into a relational database definition entails mapping every OPM object or protocol class C into a primary relation R. Depending on their type (primitive, abstract, simple, tuple, etc.), non-derived attributes of C are mapped into local attributes of R, foreign-key attributes of R, or additional auxiliary relations with appropriate foreign-key to primary-key references. Derived OPM attributes are mapped into relational procedures that are used for computing their values at run time. The OPM Schema Translator also generates a *mapping dictionary* containing information on the OPM to relational database mapping.

The OPM Retrofitting Tools

The OPM Retrofitting tools [5] can be used for constructing and maintaining OPM views on top of existing flat files or relational databases that were not developed using OPM. These tools follow an iterative strategy for constructing OPM views:

first a canonical (default) OPM view is generated automatically from the underlying database schema; then this canonical OPM view can be refined using schema restructuring operations, such as renaming or removing classes and attributes, merging and splitting classes, adding or removing subclass relationships, defining derived classes and attributes, and so on. A mapping dictionary records the information regarding the relationships between the view (OPM) constructs and their corresponding representations in the underlying database.

The OPM Database Query Tools

The OPM Database Query tools provide support for specifying and processing OPM-QL queries over native OPM databases, generated using the OPM Database Development tools, or databases retrofitted with an OPM view, and for browsing the results of these queries. In addition, for native OPM databases, the query tools support data manipulation (inserting, deleting and updating). The queries are evaluated using OPM Query Translators which employ the information in the mapping dictionary generated by the Schema Translator or Retrofitting tools in order to generate equivalent queries using the query facilities provided by the underlying DBMS or file system. For relational databases, the OPM Query Translators generate SQL queries in the particular dialect of SQL supported by the underlying relational DBMS, and then convert the query results into an OPM data structure.

Flat files are queried using SRS (Sequence Retrieval System) [8], a system originally developed at the European Bioinformatics Institute for accessing archival sequence databases. SRS parses flat files into an object structure that can be used as the initial (canonical) schema for the OPM Retrofitting Tools, and also indexes these files. The query facilities provided by SRS are limited to regular-expression searches on indexed string fields and comparisons on numeric fields, and therefore OPM queries cannot be entirely translated into SRS queries. Consequently, in order to provide general OPM query facilities on flat files, it is often necessary to perform further local processing of the SRS query results using the OPM Multidatabase Query Processor described below.

Application programs can interact with the OPM query translators either via a C++ API or by calling the query translators as Unix command-line programs. The later can be achieved using Perl or Unix shell scripts, via temporary files for passing OPM queries and query results.

The OPM Web Query Interface [6] has been designed to provide an extension to the ubiquitous Web (HTML) query forms that users are already familiar with, so that using this interface will not require learning an entirely new querying paradigm. Instead of providing predefined query forms, the OPM Web Query Interface provides support for constructing a query tree by selecting classes and attributes of interest using a graphical user interface, and for dynamically generating HTML query forms based on this query tree. Further query condition specification can be carried out by filling in these query forms.

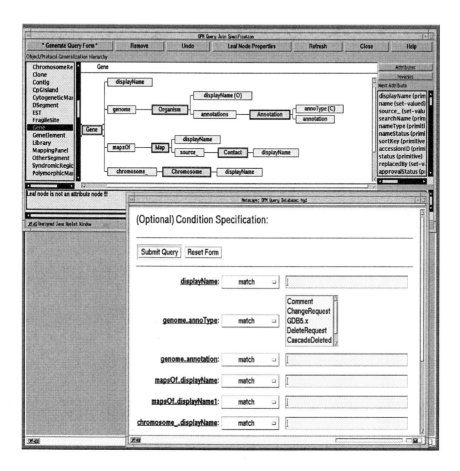

Figure 2: Constructing an OPM Query with the OPM Web Query Interface.

Query specification using the OPM Web Query Interface is illustrated by the example shown in Figure 2, where class Gene is selected as the root of the query tree. Attributes such as displayName, genome, mapsOf, and chromosome are then selected from the list of attributes associated with class Gene and added to the query tree. Next, the value classes of abstract attributes, such as chromosome, can be selected and their attributes, for example attribute displayName of class Chromosome, can be added to the query tree. Primitive attributes, such as displayName and annotation, form the leaves of the tree. Once the query

194

tree is completed, an HTML query form (see the form in the lower half of Figure 2) is generated for specifying conditions.

The OPM Multidatabase Tools

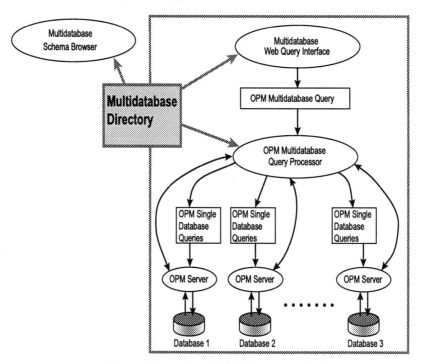

Figure 3: The OPM Multidatabase Tools

The OPM Multidatabase tools [7] provide facilities for exploring, querying and combining data from multiple heterogeneous databases via their native OPM schemas or retrofitted OPM views. The tools employ a Database Directory as described above, that records the metadata needed to access databases and information about the inter-database links.

The diagram in Figure 3 shows the architecture of the main OPM Multidatabase tools. A Java-based OPM Multidatabase Schema Browser, similar to the OPM Schema Browser for single databases described above, allows browsing the schemas of multiple databases and following inter-database links. A Web-based Multidatabase Query Interface provides support for interactively specifying OPM queries across multiple databases using a combination of graphical Java-based tools and HTML forms. This query interface is similar to the single-database OPM Web Query

Interface, except that one can first select a database from a list, before selecting the classes that will form the query tree. The top two windows in Figure 4 show an example of the Web query interface in use. In this example, first GSDB is selected from the list of component databases and then class Project of GSDB is selected as the root of the query tree. Further, the query tree may involve inter-database links in addition to regular OPM attributes, such as GSDB_to_GDB_gene in the example. These links are used to associate classes in different databases. Once the query tree is completed, an HTML query form is generated for specifying conditions, possibly involving attributes of classes from different databases (see the form in the middle part of Figure 4). The Multidatabase Web Query Interface generates queries in the OPM Multidatabase Query Language, OPM*QL, which are then executed using the OPM Multidatabase Query Processor. The bottom window in Figure 4 shows a OPM*QL query equivalent to the query expressed using the Web query tools in the other two windows.

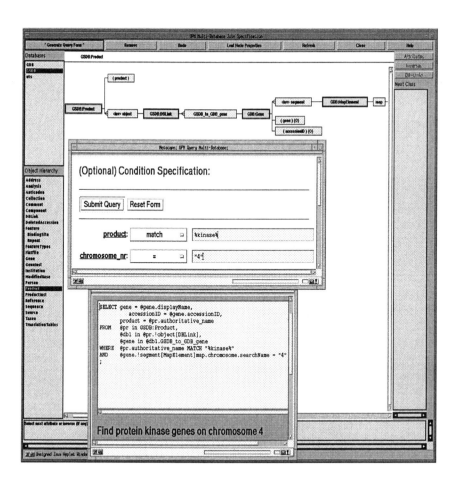

Figure 4: Constructing an OPM Query across GDB and GSDB with the OPM Multidatabase Web Query Interface.

The OPM Multidatabase Query Processor interacts with Database Servers for each database involved in the multidatabase system. Each Database Server provides DBMS specific query translation functions and facilities to execute single-database queries expressed using OPM-QL. The OPM Multidatabase Query Processor interprets OPM*QL, generates queries over the individual databases, and performs local data manipulations necessary to combine the results of individual queries and to provide functionality not supported by the individual databases. Since different databases and DBMSs may support different query facilities, the queries generated for each database are dependent on the particular subset of OPM-QL supported by the Database Server: in general as many of the query conditions as possible are performed by the Database Servers, while conditions which may not be tested by the remote DBMS are evaluated locally by the Multidatabase Query Processor.

Applications and Experience

In this section, we briefly discuss typical applications of the oPM tools and the experience we gained developing and applying these tools.

The OPM Database Development and Query Tools

The OPM Database Development Tools were first used for developing a prototype database for a large-scale DNA sequencing project at Caltech.[5] Subsequently, the OPM Database Development and Query Tools were employed for developing and then extending version 6 of the Genome Database (GDB) using the Sybase DBMS, and the Primary Database of the German Human Genome Resource Center (RZPD) using the Oracle DBMS, as well as other scientific and traditional databases. OPM Web Query Interfaces are employed for accessing several OPM based databases such as RZPD, while GDB is accessed via custom-built Web query interfaces.

OPM and the OPM tools helped in improving the efficiency of developing and maintaining these databases and, to some degree, in insulating their applications from the underlying DBMSs. As a methodology, OPM encourages and provides support for comprehensive database documentation. Various OPM tools provide facilities for taking advantage of this documentation during database exploration. Further, by insulating applications and users from the underlying DBMSs, the OPM tools simplify the task of transferring databases to other DBMSs such as object-relational DBMSs.

---

[5] See http://gizmo.lbl.gov/jopmDemo/shotgun.html

Our strategy of providing support for wrappers on top of commercial relational DBMSs has proved to be effective. In spite of early doubts expressed in the genome database community, relational DBMSs are widely used for implementing biological databases, while usage of object-oriented DBMSs has been limited and problematic (see [9] for details). The emerging object-relational (*Universal Server*) DBMSs, such as Oracle 8 and the Informix Universal Servers, which represent the next generation DBMSs, are evolving from their relational counterparts and will replace them eventually. However, while the added functionality of these new DBMSs allows the development of potentially more powerful databases, it also increases the complexity of designing, implementing and querying such databases (see chapter 15 of [12]). We believe that the OPM tools, appropriately adapted and enhanced, will continue to provide the same advantages for databases implemented with Universal Servers as those currently provided for relational databases.

Since keeping track of historical information is important for archival databases such as GDB, we have incorporated a versioning mechanism into OPM. The research underlying this mechanism, conducted in collaboration with our colleagues at GDB, lead to interesting results [4], including the realization that the implementation of such a mechanism in a relational database framework causes an unacceptable overhead for large production databases.

Our work on the OPM Database Development and Query Tools has benefited from the feedback and suggestions received from our collaborators and users. Especially valuable has been our close collaboration with the GDB staff, including Ken Fasman, Stan Letovsky, Peter Li and their colleagues. This collaboration helped us cope with the peculiarities of the Sybase DBMS and proved invaluable in improving the performance of the SQL code generated by the OPM tools. Further, GDB staff provided suggestions for extending the OPM tools with new capabilities. The more recent collaboration with the RZPD group lead by Brian Toussaint has also been instrumental in improving the OPM tools.

The OPM Retrofitting and Multidatabase Tools

We have been experimenting with the OPM Retrofitting and Multidatabase Tools since January 1996, and our experience with these is therefore more limited than with the other OPM tools. We have applied the OPM Retrofitting Tools to several relational databases, including the Genome Sequence Database (GSDB) and the bio-collections database of UC Berkeley's Museum of Vertebrae Zoology (MVZ), and to biological flat file databases such as GenBank. Retrofitting allowed us to install the OPM Web- Browsing and Query Interfaces on top of these databases. While more powerful and flexible than the native Web based interfaces provided by these databases, the OPM query interfaces may not perform as well as the canned-queries that have been manually optimized for directly accessing these databases.

Our collaboration with Thure Etzold at the European Bioinformatics Institute in installing the OPM Retrofitting Tools and Query Interfaces on top of SRS, has provided the ability to access via OPM interfaces a wide variety of structured flat-file databases, including many of the major archival molecular biology databases. SRS reads and indexes flat-files using parsers defined using in the Icarus language, and then maps them into objects. Initially, we have employed existing Icarus parsers for various molecular biology databases such as GenBank. Next, we intend to experiment with more sophisticated parser definitions and new retrofitting techniques that would allow constructing more detailed and semantically richer OPM views for such databases.

The OPM Multidatabase Tools have been applied to the construction of a Molecular Biology Database Federation that includes GDB, GSDB, and GenBank. The experience gained with this prototype has driven several enhancements of the Multidatabase Query Tools.

Implementation Issues

The OPM Schema and Query Translators have been developed mainly in C++. The OPM Schema Editor, Browser, and Query Interfaces were first developed using X11/Motif. We encountered numerous problems maintaining and porting these early OPM editors and interfaces. Currently all our OPM editors and interfaces are implemented in Java. In spite of its present instability, using Java substantially reduced the development and maintenance of these tools.

For implementing interfaces between the OPM Web Query Interfaces and Query Translators, we are using CGI, which is easy to use and maintain given our multiple-language programming environment. CGI is a natural choice for dynamically creating HTML pages. In the current CGI-based implementation, the Java-based query construction front-end uses CGI to call a Perl script wrapper that invokes a C++ version of the OPM Query Translator adapted for generating HTML files. We have also considered other communication alternatives, including Java and C++ sockets, CORBA based products with Java and C++ interfaces, such as IONA's Orbix and OrbixWeb, and Java 1.1's Remote Method Invocation (RMI) interface, with C++ applications accessed through the Java Native Interface (JNI). These alternatives are discussed in more detail in [6].

The interfaces between the OPM Multidatabase Query Processor and the OPM Database Servers are implemented using a CORBA product (either IONA's Orbix or Visigenic's VisiBroker). As a programming environment, CORBA's object-based communication between applications (possibly developed in different languages) is convenient. However the Interface Definition Language (IDL) is limited in the data structures that can be passed between applications, and therefore requires extraneous conversions between the data structures used in applications and those that can be communicated. Further, despite the C++ mapping defined in the CORBA 2.0 standard, CORBA implementations remain vendor specific, so that porting an implementation of our query tools to different CORBA products is time consuming.

Acknowledgments

Between 1992 and 1997, the OPM tool development was carried out in the framework of the Data Management R&D Group at Lawrence Berkeley National Laboratory (LBNL), with funding provided by the Office of Biological and Environmental Research and the Mathematical, Information, and Computational Sciences Division of the Office of Energy Research, U.S. Department of Energy under Contract DE-AC03-76SF00098. We want to thank Arie Shoshani, the head of the Data Management R&D Group at LBNL, for his active support and encouragement.

References

1. Chen, I. A. and Markowitz, V. M. *An Overview of the Object-Protocol Model (OPM) and OPM Data Management Tools.* Information Systems, 20(5), 1995, pp. 393-418.
2. Chen, I. A. and Markowitz, V. M. *The OPM Schema Translator.* Technical Report LBNL-33706, Lawrence Berkeley National Laboratory, 1996.
3. Chen, I.A., Kosky, A., Markowitz, V.M., and Szeto, E. *The OPM Query Translator.* Technical Report LBNL-33706, Lawrence Berkeley National Laboratory, 1996.
4. Chen, I. A., Markowitz, V.M., Letovsky, S.I., Li, P., and Fasman, K.H., *Version Management for Scientific Databases.* Advances in Database Technology- EDBT-96. Lecture Notes in Computer Science, vol. 1057, P. Apers & al (eds), Springer-Verlag, pp. 289-303, 1996.
5. Chen, I.A., Kosky, A.S., Markowitz, V.M., and Szeto, E. *Constructing and Maintaining Scientific Database Views.* Proceedings of the 9th Conference on Scientific and Statistical Database Management, IEEE Computer Society, 1997, pp. 237- 248.
6. Chen, I.A., Kosky, A.S., Markowitz, V.M., and Szeto, E. *Exploring Databases on the Web.* Technical Report LBNL-40340, Lawrence Berkeley National Laboratory, 1997.
7. Chen, I.A., Kosky, A., Markowitz, V.M., and Szeto, E., *Exploring Heterogeneous Biological Databases: Tools and Applications.* Technical Report LBNL-40728, Lawrence Berkeley National Laboratory, 1997.
8. Etzold, T., and Argo, P. SRS, An Indexing and Retrieval Tool for Flat File Data Libraries. Computer Applications of Biosciences, Vol. 9, No.1, pp. 49-57, 1993. See also hp://www.embl-heidelberg.de/srs/srsc.
9. Goodman, N. *An Object-Oriented DBMS War Story: Developing a Genome Mapping Database in C++.* In Modern Database Management: Object-Oriented and Multidatabase Techniques, W. Kim (ed), ACM Press, 1994.
10. *The Object Database Standard: ODMG-93.* Cattell, R. G. G. (ed), Morgan Kaufmann, 1996..
11. *Programmer's Reference.* National Center for Biotechnology Information, 1991. See also: http://www.inria.fr:80/rodeo/personnel/hoschka/asn1.html.
12. Stonebraker, M. *Object-Relational DBMSs: The Next Great Wave.*Morgan-Kaufman Publishers, Inc., 1996.

17 BIOKLEISLI: INTEGRATING BIOMEDICAL DATA AND ANALYSIS PACKAGES

Susan B. Davidson[*], O. Peter Buneman*,
Jonathan Crabtree*, Val Tannen*, G. Christian
Overton* and Limsoon Wong **

* Center for Bioinformatics, University of Pennsylvania,
Philadelphia, PA 19104

** Institute of Systems Science, Singapore 119597

Introduction

A vast amount of information is currently available in electronic form with Inter- and Intranet access.   The ability to use this information involves several distinct problems: first, knowing where the information is that pertains to a particular area of interest; second, accessing the information rapidly; third, efficiently integrating and potentially transforming the information into a different form; and fourth, viewing the results in an appropriate manner.

Within the Bioinformatics community, researchers typically solve the first problem by formally or informally notifying each other of the existence of various data sources through workshops, conferences, publications, registration on community web pages, etc. That is, there is general knowledge of what the various primary data sources are and some level of documentation available on how to access the data sources and retrieve information. The data is also made accessible to the community by submission of queries by email or granting remote login privileges. There are also a variety of mechanisms within the community for visualizing various types of data, contributing to a solution to the fourth problem. For example, postscript files are typically presented using ghostview or some similar displaying tool, 3-D chemical structures are typically viewed using a variety of sophisticated graphical packages (such as Rasmol), and sequence data can be viewed using a variety of tools, such as those developed using bioWidgets.

However, a fundamental barrier exists to solving the second and third problems since the format of the data and functionality of access routines for the data can vary dramatically from source to source. While commercial tools exist for combining data from multiple *relational* databases, they do not extend beyond the "sets of records" type system of relational databases to more complex types such as are found in ASN.1 and Ace formats. It is therefore difficult, if not impossible, to use a single language or access mechanism to obtain, combine and efficiently transform data from multiple non-relational sources.

BioKleisli represents a solution to this problem by providing a uniform language for querying and combining information from a wide variety of data sources, including the complex data sources that typify the biomedical research community – ASN.1, AceDB, SRS indexed files, EcoCyc and other Lisp-based reasoning systems, in addition to relational systems such as Oracle and Sybase. This list of types of data sources is not exhaustive; BioKleisli can easily be extended by adding new "data drivers" as new types of data sources are encountered. In addition to its expressive query language and extensible architecture, BioKleisli has a powerful query optimizer that generalizes many of the well-known optimizations of relational systems to this richer "complex type" system, providing significant improvements in run-time performance.

BioKleisli can be thought of as a data access, transformation and integration toolkit, as shown by the dotted lines in the figure below. It is middleware, sitting between heterogeneous data sources (shown at the bottom of the figure) and tools that operate on some integrated version of the available data (shown at the top of the figure). For example, data mining tools must first have the available data in a standard format – commonly relational – before they can be applied. With the integration and transformation capabilities of BioKleisli, users can create either (virtual) views of the underlying data sources, or instantiate data warehouses. Using a virtual approach, data is accessed at the underlying sources represented in the view. Data is always up-to-date since it is accessed at its source; however, queries on the view may encounter network delays since the data is not local. Using a warehouse approach, data is extracted once from the underlying data sources and stored in some local database. While queries on the warehouse will not encounter network delays, the data may not be completely up-to-date. The "freshness" of the data depends on the update policies used by the warehouse.

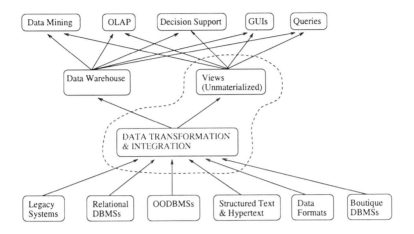

In the remainder of this chapter, we describe the complex type model underlying BioKleisli, and a query language for manipulating these types called the "Collection Programming Language" (CPL). We then briefly describe the use of BioKleisli through an example of a parameterized view. We close by summarizing how BioKleisli compares with other systems in use within bioinformatics, in particular OPM and CORBA.

The Complex Type Model

The type system underlying BioKleisli goes well beyond the "sets of records" type system of the relational model and allows complex types – arbitrarily nested records, sets, lists, bags and variants. Recall that a bag (also called a multi-set) is a set in which duplicates may occur, and that a list is a bag with order.

As an example of a complex type, consider the Publication type shown below, which has been taken from the ASN.1 citation literature (1). Note the nesting of a set of keywords (strings) within the keywd record field of the Publication type, the nesting of author records within the authors record field, and the use of a variant or "tagged union" type within the journal field representing that publications are either controlled journal entries (also a variant type), or uncontrolled entries containing the name of the person who performed the data entry.

Publication = *SET{*
   RCD{ title: *string,*
        authors: *LIST{ RCD{ name: string,* initial: *string}}*

```
journal: VRT: { uncontrolled: string,
                controlled: VRT{ medline-jta: string,
                                  \* Medline journal title abbrev.\*
                                 iso-jta: string,
                                  \*ISO journal title abbreviation\*
                                 journal-title: string,
                                  \* Full journal title\*
                                 issn: string }}
                                  \*ISSN number\*
      volume: string,
      issue: string,
      year: int,
      pages: string,
      abstract: string,
      keywd: SET{string}} }
```

In general, the types are given by the syntax:

$$\tau := int|bool|string|\ldots|SET\{\tau\}|BAG\{\tau\}|LIST\{\tau\}|VRT\{l_1{:}\tau_1,l_2{:}\tau_2,\ldots,l_n{:}\tau_n\}|$$
$$RCD\{l_1{:}\tau_1,l_2{:}\tau_2,\ldots,l_n{:}\tau_n\}$$

Here $bool\,|\,int\,|\,string\,|\ldots$ are the (built-in) base types. The other types are all *constructors* and build new types from existing types. $RCD\{l_1{:}\tau_1,l_2{:}\tau_2,\ldots,l_n{:}\tau_n\}$ constructs record types from the types $\tau_1,\tau_2,\ldots,\tau_n$. $VRT\{l_1{:}\tau_1,l_2{:}\tau_2,\ldots,l_n{:}\tau_n\}$ constructs variant types from the types $\tau_1,\tau_2,\ldots,\tau_n$. $SET\{\tau\}$, $BAG\{\tau\}$, and $LIST\{\tau\}$ respectively construct set, bag, and list types from the type τ.

Values of these types can be explicitly constructed in CPL as follows: $RCD\{l_1{:}e_1,l_2{:}e_2,\ldots,l_n{:}e_n\}$ for records; $VRT\{l:e\}$ for variants, $SET\{e_1,\ldots,e_n\}$ for sets; and similarly for bags and lists. For example, a fragment of data conforming to the Publication type is

```
SET{ RCD{ title : "Structure of the human perforin gene",
          authors : LIST{RCD{ name : "Lichtenheld",
                              initial : "MG"},
                         RCD{ name : "Podack",
                              initial : "ER"}},
          journal: VRT{ controlled: VRT{medline-jta:"J Immunol"}}
          volume:"143",
          issue:"12",
          year:1989,
```

```
        pages : "4267-4274",
        abstract: "We have cloned the human perforin gene....",
        keywd : SET{"Amino Acid Sequence", "Base Sequence", "Exons",
            "Genes, Structural"} } ...}
```

where the "…" indicates that there are other records in the set that have been omitted. Translating from ASN.1 to this format is straightforward, as it is for a variety of other data formats. We should remark here that BioKleisli does not represent entire databases in this format; it is used for data exchange between the query language of a DBMS or the application programming interface of a data format.

The Collection Programming Language (CPL)

The syntax of CPL used here is similar to that of OQL (2), the ODMG standard for object-oriented database languages. Rather than giving the complete syntax, we will illustrate it through a series of examples. The first example extracts the title and authors from a database DB of the type Publication:

setof rcd{ title : p.title, authors : p.authors}
where \p ← DB

Note the use of " \p " to introduce the variable *p*. The effect of " \p ← DB " is to bind *p* to each element of the set DB. The use of explicit variable binding is needed when queries are used in conjunction with function definition or *pattern matching* as in the example below, which is equivalent to the one above. Note that the ellipsis "…" matches any remaining fields in the DB record.

setof rcd{ title : t, authors : a}
where rcd title : \t, authors : \a, ...} ← DB

Also, the following queries are equivalent:

setof rcd { title : t, authors : a}
where rcd { title : \ t,
*   authors : \a,*
*   year : \y ... } ←DB,*
*   y = 1988*
setof rcd { title : t, authors : a}
where rcd { title : \ t,
*   authors : \ a,*
*   year : 1988, ...} ← DB*

These queries are no more than simple projection-selection queries and, but for the fact that the source data is not in first-normal-form, could be expressed in a relational query language. However, CPL can perform more complex restructurings such as nesting and unnesting, as shown in the following examples.

setof rcd { title *: t,* keyword *: k}*
where rcd { title *: \t,* keywd *: \kk, ...}* ← DB, \k ← kk

setof rcd{ keyword : *k,* titles *: setof* x.title *where* \x ← DB, k ←x.keywd*}*
where \y ← DB, \k ← y.keywd

The first query "flattens" the nested relation; the second restructures it so that the database becomes a database of keywords with associated titles. Operations such as these can be expressed in nested relational algebra and in certain object-oriented query languages. The strength of CPL is that it has more general collection types, allows function definition and can also exploit variants, which may be used in pattern matching:

setof rcd{ name *: n,* title *: t }*
where rcd{ title *: \t,* journal *: vrt{* uncontrolled *: \n}, ...}* ← DB

This gives us the names of "uncontrolled" journals together with their titles. The pattern "*vrt{* uncontrolled *: \n}*" matches only uncontrolled journals and, when it does, binds the variable *n* to the name.

The syntax of functions is given by \x =*e*, where *e* is an expression that may contain the variable *x*. We can give this function (or any other CPL expression) a name with the syntax *define f (\x)=e*, which causes *f* to act as synonym for the expression *e*. Thus, the titles of papers of a given author can be expressed as the function,

define papers_of (\x) = *setof p where p* ← DB, *x* ← *p*.authors

Note that \x ← *p*.authors matches elements of a list rather than elements of a set.

These examples illustrate part of the expressive power of CPL. A more detailed description of the language is given in (3), where a description of how to express aggregate functions such as summation, as well as functions such as transitive closure, is also given. CPL can also be used to query object-oriented databases by including a reference type, a reference pattern, and a dereferencing operation (4).

BioKleisli in Action: Querying Biomedical Databases

BioKleisli consists of a query execution engine and a set of type specific data drivers. The figure below illustrates the installation that we use at the Center for Bioinformatics (PennCBI) which underlies the queries shown at the CPL website http://www.pcbi.upenn.edu (follow links to research projects). Users interact with the PennCBI installation using parameterized HTML query forms, a variety of programs written in perl5, prolog, C and other languages, or by using CPL directly. The types of external data sources that we connect to include ASN.1, Sybase and AceDB, as well as the BLAST sequence analysis package, as shown by the types of data drivers illustrated. Note that the Sybase driver can be used for both the GDB Sybase server as well as our local Sybase database for Chromosome 22, Chr22DB.

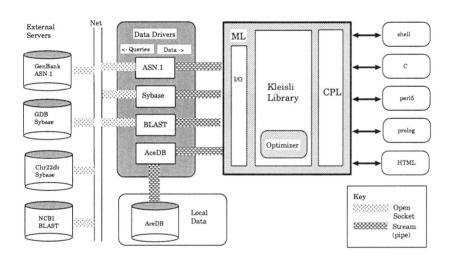

To query an external server, the names and types of "structures" that will be accessed in the data source must be registered as primitives in the BioKleisli library. For example, with a relational database one must register as parameterless functions the names of relations or views that will be accessed; alternatively, one can simply register one function per database which takes as input the name of a relation and returns a result of type set of records. These functions can then be used in CPL queries or within other CPL function definitions to create "user views" of the underlying data sources. The query execution module within BioKleisli will then generate the appropriate query in the host language of the external server to extract the value of the named structure. For example, with a Sybase server an SQL query would be generated from the CPL query. The host language query is then passed to the appropriate data driver, and from there to the external server. When the external server returns the result, the data driver translates it into internal BioKleisli format and returns the translated result to the query execution module for further processing within the original CPL query.

Non-Human Homolog Search

To illustrate how BioKleisli executes queries, we will walk through an example: "Find information on the known DNA sequences on human chromosome 22, as well as information on homologous sequences from other organisms." The strategy taken in writing this query will be to combine information from relational GDB and ASN.1 GenBank. GDB is queried for information about the accession numbers of DNA sequences known to be within chromosome 22. The NA-Homolog-Summary function available in the Entrez interface to ASN.1 GenBank is then invoked to retrieve homologous sequences (i.e., sequences with significant similarity to the original). The homologous sequences are then filtered to retrieve only non-human entries. The final answer is printed as a nested relation.

The GDB Query.

The GDB query joins three tables – locus, object_genbank_eref, and the portion of the locus_cyto_location that corresponds to entries on Chromosome 22– over the locus_symbol field, and projects over the locus_symbol and genbank_ref fields. Assuming that the function GDB has been registered within BioKleisli to access the contents of a table whose name must be specified by the query writer, the query is simply written in CPL as

```
define Loci22 =
setof rcd { locus_symbol :x, genbank_ref : y}
where    rcd{ locus_symbol : \x, locus_id :\a,...}  ←GDB("locus"),
         rcd{genbank_ref  :  \y,  object_id  :  a,  object_class_key:  1,  ...}
         ←GDB("object_genbank_eref"),
         rcd{loc_cyto_chrom_num:  "22",  locus_cyto_location_id  :  a,  ...}
         ←GDB("locus_cyto_location")
```

Note that we could also have written this query by registering separate functions for each table accessed (locus, locus_cyto_location and locus_symbol) and that this would have given query writers an idea of the names available within GDB.

If executed as written, Loci22 would generate three separate SQL queries to GDB, each of which would extract the contents of a table. The optimizer, however, improves this by writing the entire function as a single SQL query. In fact, a feature of the optimizer is that it is capable of moving the largest possible subquery of a CPL query to an external server for execution. This can be done for a wide variety of types of data sources (5), including relational and ASN.1-Entrez data sources. The savings in execution time is significant: Not only does it minimize the amount of data shipped from an external data source to BioKleisli, but it can take advantage of the powerful optimizers present in commercial database systems.

Revisiting the non-human homolog search query.

From the accession numbers returned by Loci22, homologous sequences can be found via the Entrez function NA-Homolog-Summary. NA-Homolog-Summary takes a DNA sequence accession number and returns a set of records describing linked entries. The final solution to our query can then be expressed using these functions as:

setof rcd{ locus *: locus,*
　　　　homologs: *setof homolog*
　　　　　　　where \homolog ←
　　　　　　　　　　　　NA-Homolog-Summary(*locus*.genbank_ref),
　　　　　　　　　not (homolog.title *like* '%human%'),
　　　　　　　　　not (homolog.title *like* '%sapiens%')}*
where \locus ←Loci22

Note that the query itself is quite simple, and that most of the effort was spent figuring out where the relevant data was stored. Examples of other queries can be found on the CPL home page at http://www.pcbi.upenn.edu (follow links to research projects) and http://corona.iss.nus.sg:8080/demos.

Conclusions

BioKleisli is a powerful toolkit – middleware – for creating data warehouses or views that integrate multiple heterogeneous data sources and software systems. Based on a complex type system in which records, variants, sets, lists and bags can be arbitrarily nested, it consists of an OQL-like query language called CPL, a query execution engine with an optimizer that extends to this complex type system many of the known optimizations of relational database systems, and a set of generic drivers which provide the translation interface to external data sources – relational, object-oriented, ASN.1, AceDB, and Lisp-based systems among others.

It is important to remember that the BioKleisli drivers for database systems are *generic.* That is, once a driver for a database management system (DBMS) has been constructed, any database that is implemented in that DBMS can be immediately opened and queried in BioKleisli. Thus, BioKleisli completely removes from the programmer the need to write any DBMS specific code. The type system used in the DBMSs and other data sources can all be naturally represented within BioKleisli, which contains drivers for a number of widely used biological data sources. With the drivers in place, the database designer may choose an integration view and express that view in CPL. The integrated view is made available to database programmers who will use the application by registering functions within BioKleisli corresponding to the structures in the integration view. For example, an integrated view for the non-human homolog search application could be one in which Loci22 and NA-Homolog-Summary, were registered as structures (the last one being a function); another could be one in which the final query was registered with some appropriate name. Note that this is a rather trivial example; in general, the integrated views will contain many structures and be much more generic.

With the increasing popularity of CORBA as a standard for software and data sharing (http://www.omg.org/library/public-doclist.html), it is worth considering the relationship between BioKleisli and CORBA. They are similar in that they both define a type system and a standard syntax/format/protocol for trafficking in values adhering to that type system. However, there are some important distinctions: BioKleisli is based around a query system, and defines a rich language for collection types. Central to the system are rewrite rules and optimizations that improve the performance of queries, and isolate portions of queries that can be locally executed by the external data sources. The CORBA specification was not written with the idea of generic optimizations and rewrite rules in mind, although individual CORBA implementations may try to do various things to improve performance.

However, CORBA is a powerful set of standards for interoperation that could be used from within BioKleisli as an external data source. That is, a CORBA driver could be written for BioKleisli, allowing BioKleisli to query any CORBA data source in addition to those it already supports. BioKleisli's type sytem is sufficiently rich to encode everything expressible in IDL. CORBA could also be used to provide a programmatic API to the BioKleisli system, allowing programmers to execute CPL queries from within a programming language of their choice. CORBA could also be used "internally" as a replacement for the current mechanism used by BioKleisli to communicate between components of the system, namely the execution engine and the data drivers, as is currently done in OPM.

OPM (6) is another integration toolkit that is popular within the Bioinformatics community. The primary difference between OPM and BioKleisli is the goal from which each project started: OPM focused on using a simple object model for presenting good visual interfaces to the user through which the underlying system(s) could be understood and queried. The original application was to retrofit relational databases with a more intuitive object model. BioKliesli focused on finding a complete language for complex types, and rewrite rules for optimizations. The type system and language underlying OPM is therefore not as rich as that underlying BioKleisli, nor is it as easy to add new data sources to OPM as it is to BioKleisli; however, the visual interface to OPM is much richer than that currently available in BioKleisli. An ideal system would combine the interfaces used within OPM with the query engine of BioKleisli; work in this area is underway.

References

13. NCBI ASN.1 Specification (Revision 2.0). Technical Report available from the National Center for Biotechnology Information, National Library of Medicine, Bethesda, MD (1992).

14. R. Cattell, D. Barry, D. Bartels, M. Berler, J. Eastman, S. Gamerman, D. Jordan, A. Springer, H. Strickland, and D. Wade. The Object Database Standard: ODMG 2.0. Morgan Kaufmann, 1996.

15. P. Buneman, L. Libkin, D. Suciu, V. Tannen, and L. Wong. *Comprehension Syntax*. SIGMOD Record, 23(1):87-96, March 1994.

16. S.B. Davidson, C. Hara and L. Popa. *Querying an Object-Oriented Database Using CPL*. Proceedings of the Brazilian Symposium on Databases, October 1997.

17. L. Wong. Querying Nested Collections. Ph.D. thesis, Department of Computer and Information Science, University of Pennsylvania, Philadelphia, PA 19104, August 1994. (Available as University of Pennsylvania IRCS Report 94-09.)

18. I.A. Chen, A. Kosky, V.M. Markowitz, E. Szeto. *OPM*QS: The Object-Protocol Model Multidatabase Query System*. Technical Report LBNL-38181, Lawrence Berkeley Laboratory, Berkeley, California, 1995. (See also http://gizmo.lbl.gov/DM_TOOLS/OPM/OPM.html.)

18 SRS: ANALYZING AND USING DATA FROM HETEROGENOUS TEXTUAL DATABANKS

Phil Carter, Thierry Coupaye, David P. Kreil, and Thure Etzold

EMBL Outstation, The European Bioinformatics Institute, Wellcome Trust Genome Campus, Hinxton, Cambridge CB101SD, UK

Introduction

Bioinformatics is a general term that may be defined as the application of computers and databases to help store, retrieve and analyze biological information. With an unprecedented growth in the quantity and diversity of biological information, bioinformatics has become a new scientific discipline widely recognized to be an integral part of future successes within molecular biology. Information growth is further complicated by independent organizations maintaining biological databases using differing technologies in unrelated ways. This has led to a multitude of dissimilar biological databanks.

SRS was initially developed as a Sequence Retrieval System to overcome these problems, but has expanded far beyond these limits today. Originally devised at the European Molecular Biology Laboratory (EMBL) in 1989, SRS's progress has continued at the European Bioinformatics Institute (EBI). SRS has evolved through many distinct stages of development, and is currently at version 5.1.0.

One fundamental property underlies most biological databases, their availability in ASCII text format (American Standard Code for Information Interchange). SRS exploits this universal medium to create an homogenous collection of databases accessible to the user through a unified interface. Furthermore, recent additions to SRS's functionality allow further analysis of retrieved information using various bioinformatics tools.

At present, SRS envelopes approximately 250 databanks worldwide at 35 public sites. Around 10,000 accesses to SRS are made each day, making it the most popular bioinformatics service at the EMBL-EBI ever.

SRS Core Features

In this section, we introduce the core features of the SRS system. SRS allows simultaneous access to different databanks and can create linked (cross-referenced) data. It is based on a parsing and indexing mechanism, to extract data from databanks and link entries from those databanks. The user can then access the data and perform complex queries across different databanks interconnected by link indices through the SRS Query Language.

Parsing and Indexing

Efficient query systems often use indices to speed up data access. SRS has such an indexing mechanism. It treats databanks as sequences of entries. Each entry is in turn composed of different data fields. The contents of the data fields are parsed (i.e. the text is broken down into recognized strings of characters for further analysis), selected words or *tokens* are isolated and then inserted into an index. There is in general a separate index for each data field.

SRS is based on Icarus [1], which refers to both the language used to extract the data by parsing the entries and the parser of this language. The Icarus parser is an essential part of SRS: the more flexible and powerful the parser is, the more information that can be extracted from the databanks. The problem of the retrieval of all available information contained in a databank is not a trivial task. Most major biological databanks, such as the EMBL [2] databank, GenBank [3], etc. were designed for human readability and are stored in many different formats. In other words, molecular biology databank formats allow significant freedom to present information that can be easily recognized by human intellect. It can be a hard problem to describe these formats using computer languages usually used for parsing such as Perl or Awk. The other problem lies in the constant and frequent revisions of databank formats which make "hard" (i.e. inflexible) coding inefficient to support the highly variable world of molecular biology databanks. The optimal solution to these problems is a special parser that allows the description of the respective formats of all known biological databanks. Within this parser the external representation of data in a flat file is converted to an internal representation, which we call a 'token list'. In contrast to most other computer language compilers (e.g. YACC [4] and Lex [5]), the Icarus parser is an interpreter and combines lexical and syntactical definitions. Both of these features significantly simplify parser programming and format descriptions of biological databanks.

In SRS, databank structures are represented by Backus-Naur Form (as described by Wirth [6]) abstract syntax trees or *grammars*. BNF grammatical rules consist of *terminal* and *non-terminal* definitions. A non-terminal represents a rule and can in turn be subdivided into terminals and non-terminals. Terminals and non-terminals group together to form a *production*. We call a piece of input stream corresponding to a single terminal a *token* and a set of tokens a *token-list*. Biological databank structures generally use the following scheme: a databank consists of a sequence of entries, an entry is made of data-fields, a data-field consists of tokens. Tokens are the part of the input that are parsed by terminals or non-terminals. For instance, a parser for a databank in an EMBL-like format might look like:

```
databank          = {entry}
entry             = id_line
                    {ra_line de_line oc_line}
                    end_line
id_line   = 'ID' id
de_line   = {'DE' word_list}
...
end_line = '//'
word_list         = {word}
word              = /[a-z]+/
```

This might operate on a text entry of the (EMBL-like) form:

```
ID  HS404281   standard; DNA; HUM; 186 BP.
DE  Human tumor suppressor (p53) gene, exon 3
OC  Eukaryota; Metazoa; Chordata; Vertebrata
OC  Mammalia; Eutheria; Primates
RA  Herrmann M., El-Maghrabi R. E., Abumrad N.N.
//
```

This grammar defines the databank as being a list of entries. The first line of each entry should be the ID-line followed by zero or more data lines, and the last line is the string '//'. The non-terminals id_line, de_line and oc_line can be recursively parsed and finally indexed.

Icarus parsers not only *recognize* entries, but can also perform some semantic actions during the parsing process. To produce output, any terminal or non-terminal can be associated with one or more different action commands, such as create a token, extend (add text to) existing tokens, set some global states of the parsing process, input/output directives, print commands, variable assignments and function calls.

Parsers generally *parse all they can*, i.e. they decompose the input starting with the root production and go on recursively until having only terminals. In SRS, this scheme is referred to as *forced parsing*. This is in opposition to *lazy parsing*, in which the parser only parses the production(s) that it is asked to parse. The example can be modified so that, if you asked for the token "entry", the parser would only go through the productions it needs, i.e. "databank" and "entry", and would not parse "id_line", "de_line", etc..

SRS is a flexible environment in which databank structures can be fully described and thus in which databanks can easily be added or changed. Nevertheless, in a system with many databanks, many indices (at least one per data-field of each databank) have to be created and updated. The link indices add substantially to the complexity of the maintenance of a SRS system since they manifest interdependencies between separate databanks. Fortunately, the index building process is automated. A program runs automatically at frequent intervals to check if a new version of a databank exists and performs the appropriate actions. The storage size for indices is relatively small, about 10 to 20% of the size of the actual indexed databanks.

Querying and Linking

On top of Icarus, a set of programs and C-API (Application Programming Interface) functions allow the interrogation of the internal Icarus representation which is superimposed to the real textual structure of databanks. Queries are expressed in the SRS Query Language which has been especially designed for the interrogation of interrelated flat file databanks. SRS queries operate on sets of entries or subentries, belonging to one or more databanks. The sets are combined using logical operators AND ('&'), OR ('|') and BUTNOT ('!'), and also two 'link-operators' denoted by the symbols '<' and '>'.

Operands include index searches of the format "[databank(s)-field:value]" where one or more databanks can be listed. For querying purposes, the sub-entries (e.g. features of a sequence) of a databank are considered as a separate databank, and can be included in this list. Queries on subentry databanks result in sets of subentries. "Field" identifies the indexed field where the search has to be performed. In the case of multiple databanks, the field must be defined for all databanks. "Value" can be a string query with wild cards ('*' and '?'), a regular expression (delimited by two forward slashes '/' at the beginning and end and using '*', '+', '?', (...), [...], etc.). Slightly different syntaxes allow numeric range or date queries. Here are some examples of queries:

Query	Meaning
[embl-key:insulin]	Simple field search
[{swissprotswissnewsptrembl} des:kinase]	multiple databanks search
[swissprot_features-Ftkey:transmem*]	subentries
[embl-des:(acetylchol*&receptor)!muscarinic]	string search
[imgt-sl#400:500]	numeric range search: between 400 and 500 bp.s
[swissprot-date#1-jan-1995:31-dec-1995]	date search
[embl-des:/^nif[a-e]$]/]	regular expression search: gene names nifa,nifb, nifc, nifd, nife

Operands also include named sets, which can either be databank names (all entries from the databank of the given name) or results of previously evaluated queries. For instance the query:

```
(temp=[embl-org:human])>SWISSPROT | temp > SWISSNEW
```

defines a temporary set named "temp" and links it to two databanks named SWISSPROT [7] and SWISSNEW.

If A and B are sets of entries of two linked databanks, the query "A>B" will return all the entries in the set B, which are linked to one or more entries in A. The link operation is antisymmetric, i.e. "A>B" is equivalent to "B<A". For sets of subentries, a predefined linking operand *parent* is defined which identifies the set of entries containing the subentries. The query "S>parent" will return all databank entries containing the subentries in S. Note that each different databank contains usually different types of data. In a certain way each databank behaves as an extent of a class in an object oriented databases. The link between two databanks has a different meaning depending on the databanks involved. For example, the query "SWISSPROT<PDB" will retrieve all known protein sequence entries (SWISS-PROT can be thought as the extent of known protein sequences) that have a known three-dimensional structure (PDB [8] contains all known protein 3d structures). Thus, the link between SWISS-PROT and PDB encodes the relationship "has a known 3d structure". Analogously, the link between a SWISS-PROT entry and EMBL entries can represent the relationship "is encoded by the DNA fragments". The link operation in SRS is reminiscent of the 'join' operator in the algebra of relational

database management systems. The join operator is used to logically combine tuples from two relations into single tuples, however the link operators select entries from one databank that are linked to entries of another databank. Link operations are much less CPU-intensive than joins especially if one of the two linked databanks is very large.

SRS Interfaces

Several different interfaces to SRS are available. The C-API facilitates the full use of SRS's capabilities to those interested in accessing function libraries for programming purposes. The second interface is a UNIX based command-line interface named 'getz'. Getz provides a more direct querying mechanism, and also implicitly allows its combination with other computer applications and UNIX commands. This allows powerful retrieval on a larger scale via automation as well as interfacing with other bioinformatics programs. The SRS Web interface is, however, by far the most popular interface in use. Although less powerful, it's user-friendly design has made SRS accessible to a much wider section of the scientific community.

SRSWWW – The Web Interface to SRS

A major difference between SRSWWW [9] and most Web services is that it maintains state [10]. In effect, users start a session within SRSWWW, at which point they are designated a unique identifier by the SRS server. After the session is completed, it may be 'bookmarked' within the Web browser. This allows the return of the user at a later time to the same point in the previous session. At present, the session is removed from the system after two days if it is not accessed again. This will be extended to an unlimited time period in the next SRS release due to increasing request for truly permanent sessions. This will also allow sessions to be downloaded and moved between different sites.

After entering a new SRS session the user enters the 'Top Page' (see figure 1). Initially, the user is requested to select at least one of the databanks offered which are grouped into categories. In addition to the inclusion of biological databases such as genome, mapping and transcription factor databases, more diverse areas such as mutation databanks are now accessible through SRS.

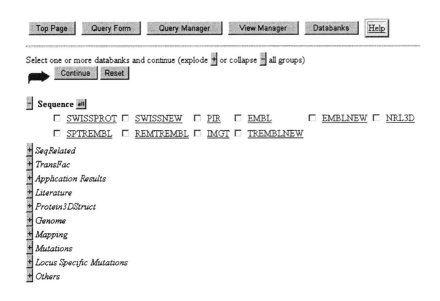

Figure 1. The SRS Top Page.

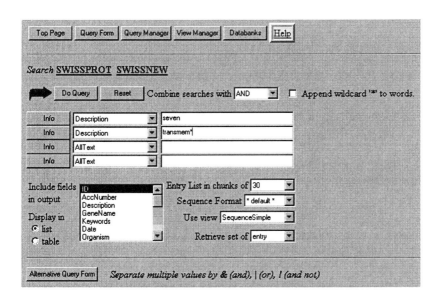

Figure 2. The SRS Query Form.

Following the selection of databanks by the user, the SRS session is continued by entering the 'Query Form' page (see figure 2). In actual fact, SRS provides two alternative query forms. The default is a simpler, hence more user friendly format. The second gives more explicit options for more detailed querying. SRS querying operates by searching a databank through its fields for given terms or expressions. The query form allows easy access to this mechanism by giving the user a choice of fields to search in their chosen database/s, into which search terms or expressions can be entered. If more than one database is chosen, only common fields between databanks are displayed. The user may also decide at this point how a result shall be displayed after retrieval, i.e. the number of entries to be retrieved at a time, the sequence format and which 'view' to use. Views are a distinct feature of SRS [11] and will be described subsequently in more detail.

Successful querying of the database/s leads to display of the 'Entry List' page. The user now has several choices as to how they will proceed. Hypertext links are inserted into the result set to allow browsing of each entire entry individually. Wherever possible, hypertext links are inserted into entries to represent cross-referencing between databanks, allowing the user to explore various depths of related information. Entire result sets or selected entries may be linked to other databases using SRS links. Several options are provided for this linking. The user may find the entries in the results set that cross-reference one or more databanks, the entries in the selected databank/s referenced by a result set, or entries in the current query results not referenced by the databanks. This can be an extremely valuable feature when used in the appropriate way, for example, a search of SWISS-PROT for all entries relating to calcium binding sites which are then linked to the PDB database. This equates to the biological question: "Give me all calcium binding proteins with a solved tertiary structure".

Another aspect of SRSWWW is its 'Query Manager'. As different queries and their results are stored at the SRS server, they can be further explored by combination with other result sets using the SRS Query Language. For example, an original query pertaining to the terms "mycobacteria" and "essential" (i.e. retrieve all essential genes found within mycobacterium), from which a result set from the same database for the keyword "operon" may be subtracted. The resulting list of entries generated by the Query Manager will therefore address the question: "All essential genes within mycobacterium not found within operons".

A last, but very useful choice, is the further analysis of the result set using various bioinformatics applications. Tools available differ between sites but approximately a total number of twenty are thought to be in use at present. This is probably one of SRS's most under exploited assets and will be further developed in its coming versions.

Analyzing Data in SRS with Applications

Besides browsing query results and using direct and indirect links between databank entries, external application programs can be used to analyze data. Typically such applications are database searches by sequence similarity (e.g., BLAST [12]), construction of multiple sequence alignments (e.g., CLUSTALW [13]), restriction map analysis, and tools predicting various properties (transmembrane regions, secondary structure, etc.).

Any system that wishes to integrate a range of applications needs to address several complex issues. These include generating a simple to use yet extensive user interface for each application, passing parameters to the application as well as supplying its input data in a suitable format, and presenting the application's results to the user.

Application Parameters and Launching

SRS currently supports two data types: nucleotide sequences and amino acid sequences. The system can in principle be extended to deal with any kind of data, be it sequence alignments, sequence profiles, or even entries in a queue of dispatched jobs. This is an area under active development.

Each entry in a set of query or link results may contain a particular data type (or several), which could be accepted as input by certain applications. The menu next to the 'launch' button on the Entry List page lists the applications that can operate on the data types present. Selecting an application and invoking the launch button leads to the respective 'Application Launch' page (see figure 3). Here one may edit the selection of data passed to the program and set a range of parameters particular to that application. By clicking on highlighted terms, context sensitive help can be requested. Help is shown in one separate dedicated window, so as not to distract the user from the main input form.

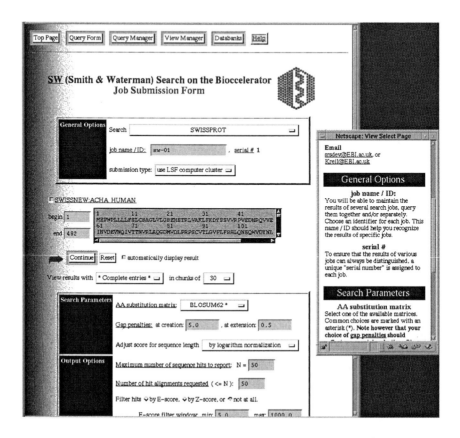

Figure 3. Key Features of the SRS interface to integrated application programs.

After submitting one's choices, the selected data and all the parameter settings made are automatically converted to a format understood by the particular application, which is then launched. In taking care of all necessary format conversions, SRS thus transparently mediates between the databases and the range of formats a data type could be stored in, and the multitude of different input formats expected by the various applications.

Processing and Viewing Application Results

Application results text is processed *like any other databank text*. This means that the results can be queried like any databank in SRS. Moreover, one can perform links to other databanks, and define specific views on application results. For example: the default view for a BLAST search links all BLAST hits to the databank searched (e.g. SWISS-PROT), and displays the description field from there. Since application results from further application launches are accumulated, one can answer questions such as "Which hits were reported by the last search but not by the previous search?", or "In the last three searches, which hits had a bit-score above 30?" The comparison is done using a link operation. This method can also be used to collate different algorithms' performance on a particular problem (e.g., a specific search by both BLAST and FASTA, or the effects of different parameters settings on an algorithm's performance).

As soon as the application program has finished running, its results are thus 'just another databank', which are consequently parsed and indexed. Each application therefore has an associated SRS databank parser, database structure description and documentation. This documentation is augmented by the help text for application parameters.

SRS Views

In the past within SRS, the retrieved information was always a set of entries each from a single databank. The specification of views was added so that the user can define virtual entries that may include information from many sources. Views can be used to treat the sum of all installed databanks as a single structure where user defined entities can freely span databank boundaries.

SRS regards an entry as a list of data-fields. The simplest form of a view acts as a filter and allows only selected fields to be displayed. For instance, a view could be defined to display only the ID, the description and the line containing the sequence length for SWISS-PROT entries as shown in figure 4. This type of view, the 'list-view', displays the selected data-fields in the order they occur within the entry and in their original format which includes line codes or field labels. It is mostly useful for 'turning off' undesired information during browsing large sets of entries.

```
ID    ACHA_HUMAN      STANDARD;      PRT;    482 AA.
DE    ACETYLCHOLINE RECEPTOR PROTEIN, ALPHA CHAIN PRECURSOR.
SQ    SEQUENCE    482 AA;   54545 MW;   E050B513 CRC32;
```

Figure 4. A SWISS-PROT entry displayed by a list-view.

The table-view provides more sophistication and possibilities. The entry in the

ACHA_BOVIN	ACETYLCHOLINE RECEPTOR PROTEIN, ALPHA CHAIN PRECURSOR.	457
ACHA_CHICK	ACETYLCHOLINE RECEPTOR PROTEIN, ALPHA CHAIN PRECURSOR.	456
ACHA_ELEEL	ACETYLCHOLINE RECEPTOR PROTEIN, ALPHA CHAIN (FRAGMENT).	24

Figure 5. A list of SWISS-PROT entries shown in a table-view.

table-view is displayed as a single row. The information displayed is in a different format in the sense that it is shown without the field identification (line codes or field labels). On top of this, other conversions can take place to render the information into a more standardized format, e.g. all author names from all data-banks can be converted into the form surname-comma-initials (e.g., "Jones, D.G."). The extraction and conversion of the information is accomplished by the Icarus parser. Figure 5 shows the same information as in figure 4 using a table view.

A view within SRS is independent from queries and can be applied to any set of entries provided they are from the databank for which the view was specified. It is possible to specify a view for multiple databanks so that a single view can be applied to sets generated by querying several databanks at the same time. However, one needs to be careful since particular data fields may not be shared by all databanks for which the view is defined.

Views become particularly useful when combined with links. Each entry displayed can be individually linked with entries from other databanks from which again only selected information is displayed. Since the linked entries in turn can be displayed by a view that specifies further links, the overall structure can be seen as a tree. The databanks for which the view is defined are root databanks and the databanks supplying the linked entries leaf-databanks. Figure 6 shows SWISS-PROT entries together with linked entries from ENZYME [14].

SWISSPROT	Description	ENZYME	CatalyticActivity
MDH_HALMA	MALATE DEHYDROGENASE (EC 1.1.1.37).	1.1.1.37	(S)-MALATE + NAD(+) = OXALOACETATE + NADH.
ODPA_ASCSU	PYRUVATE DEHYDROGENASE E1 COMPONENT, ALPHA SUBUNIT, TYPE I PRECURSOR (EC 1.2.4.1) (PDHE1-A).	1.2.4.1	PYRUVATE + LIPOAMIDE = S-ACETYLDIHYDROLIPOAMIDE + CO(2).

Figure 6. Two SWISS-PROT entries shown with the Description field together with their linked entries from ENZYME for which the "Catalytic Activity" field is included.

An individual data-field may have a list of format options. The sequence field of protein sequence databanks for example, can be displayed as the plain sequence of characters, in GCG, PIR or FASTA format. Another option is its display as a Java applet, which shows the sequence along with a plot for various amino acid characteristics such as the Kyte-Doolittle [15] hydropathicity values.

Figure 7 shows a SWISS-PROT entry with the sequence displayed by a Java applet. Displaying Java applets within views is an elegant way of linking data with analysis methods and is straightforward, since HTML provides tags for calling applets and supplying their input.

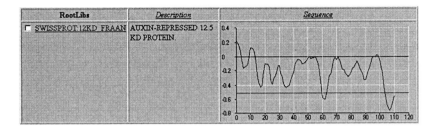

RootLibs	Description	Sequence
☐ SWISSPROT 12KD FRAAN	AUXIN-REPRESSED 12.5 KD PROTEIN.	

Figure 7. A SWISS-PROT entry with its sequence shown by a Java applet as a hydropathicity plot. The applet can be controlled to display other amino acid properties or to show the curve with a larger scale.

Since a single root entry may be linked to many leaf entries it is sometimes useful to have only the number of linked entries displayed. The SRSWWW server shows this number as a hypertext link that when clicked retrieves the list of leaf entries. The number is of value itself and can be used for statistics. For instance, using an example of a link from the TAXONOMY [16] databank to GENBANK one could obtain a table of taxa together with the number of GENBANK entries that exist for all organisms each taxon represents.

SRS World Wide: The Databank of DATABANKS

Locating an appropriate server is only the first problem users face when exploring new databases, since there are as many different query interfaces as there are databanks. Documentation relating to the databanks and manuals describing the interfaces vary greatly in content and form, and are often not easily obtained.

In this section we focus on a newly added component within SRS version 5.1 [17] which generates DATABANKS, a database of databanks, by traversing public SRS servers around the world. In SRS, the documentation for each databank can be viewed as the respective 'databank information page' via the SRSWWW interface. The databank information page has a standardized layout for easy reference and also includes server status data on indexing of the respective databank. Both the parser and the databank documentation files can be requested from SRS for reuse at other sites. By pooling the knowledge gathered at these worldwide sites, SRS now provides both a unified method for direct access to around 250 different databases and a standard framework for their documentation. Automated nightly compilation of the data guarantees that the overview is up to date.

The documentation collected in DATABANKS is written by the community of SRS server administrators. In a collaborative effort, it is regularly updated and extended in a distributed manner.

New Functionality for the User

DATABANKS gives the user a single uniform entry point for browsing or searching particular databanks, and optionally leads directly to the respective databank query forms. DATABANKS is typically searched by databank name or description. More generally, any fields of the databank information pages as well as site and server characteristics can be used in a query.

The results of such a search show all the databanks that matched the search request and at which sites they are available. For convenience, the list of results offers direct links to their remote SRS query forms. For example, figure 8 shows the result of a request for databanks named 'ENZYME'. As in most cases, more than one server maintains a copy of the databank, and the list shows alternative sites. The number of indexed entries and the release number (where assigned by the server maintainers) help in choosing a nearby site with a current version of the databank. When searching for a particular database, it often helps to first restrict the search to a subset of DATABANKS which includes only one site from each group of alternatives. Currently, this representative site is chosen as the site with the most extensive databank information page.

DATABANKS	Release	EntriesN	IndexDate	RemoteLinks
⌐DATABANKS:ENZYME at IBMM–DBM, Université Libre, Brussels, Belgium	22.0	3650	22–Jan–1998	Query / Index
⌐DATABANKS:ENZYME at Belgian EMBnet Node (BEN), Brussels, Belgium	22.0	3650	23–Feb–1998	Query / Index
⌐DATABANKS:ENZYME at CSC, Otaniemi, Espoo, Finland	20.0	3601	05–Nov–1997	Query / Index
⌐DATABANKS:ENZYME at DKFZ, Heidelberg, Germany	22	3650	16–Jan–1998	Query / Index
⌐DATABANKS:ENZYME at HGMP–RC, Hinxton, Cambridge, UK	21.0	3650	23–Dec–1997	Query / Index
⌐DATABANKS:ENZYME at IUBio, Indiana University, USA		3645	14–Nov–1997	Query / Index
⌐DATABANKS:ENZYME at EMBL–EBI, Hinxton, Cambridge, UK		3650	10–Mar–1998	Query / Index
⌐DATABANKS:ENZYME at GBF, Braunschweig, Germany	22.0	3650	29–Jan–1998	Query / Index
⌐DATABANKS:ENZYME at Vienna Biocenter EMBnet Node, Vienna, Austria	22.0	3650	06–Feb–1998	Query / Index
⌐DATABANKS:ENZYME at CAOS/CAMM Center, Nijmegen, The Netherlands	22.0	3650	03–Feb–1998	Query / Index
⌐DATABANKS:ENZYME at CBI EMBnet Node, University of Beijing, China		3645	03–Dec–1997	Query / Index
⌐DATABANKS:ENZYME at EMBL, Heidelberg, Germany		3650	30–Nov–1997	Query / Index
⌐DATABANKS:ENZYME at ExPASy, Geneva, Switzerland		3650	07–Mar–1998	Query / Index
⌐DATABANKS:ENZYME at IBB–PAS EMBnet Node, Warsaw, Poland		3601	28–Jan–1998	Query / Index
⌐DATABANKS:ENZYME at INFOBIOGEN, Villejuif, France	22.0	3650	06–Jan–1998	Query / Index
⌐DATABANKS:ENZYME at MBDC Oxford, Oxford University, UK		3645	17–Oct–1997	Query / Index
⌐DATABANKS:ENZYME at Institut Pasteur, Paris, France		3650	05–Dec–1997	Query / Index
⌐DATABANKS:ENZYME at Sanger Centre, Hinxton, Cambridge, UK	22.0	3650	04–Dec–1997	Query / Index
⌐DATABANKS:ENZYME at SEQNET EMBnet Node, Daresbury, UK	22.0	3650	19–Jan–1998	Query / Index
⌐DATABANKS:ENZYME at IVR, Kyoto University, Japan		3645	06–Mar–1998	Query / Index

Figure 8. The results of a query for databanks named 'ENZYME'.

Each entry in DATABANKS contains a copy of the remote SRS databank information page, which includes field descriptions and data about the indices, and it concludes with an overview of alternate sites. Parts of a typical entry are shown in figure 9. The overview provides direct links for remote queries to each of the sites. For users in the network vicinity of a particular DATABANKS server, the relative response times compiled by that server give a clue of the net distances to other sites. If problems are encountered with the connection to a particular site, it is moved to the end of the list of alternatives. In this case, data from previous runs is used as backup. A record of when the backup was originally retrieved indicates whether it might be out of date.

228

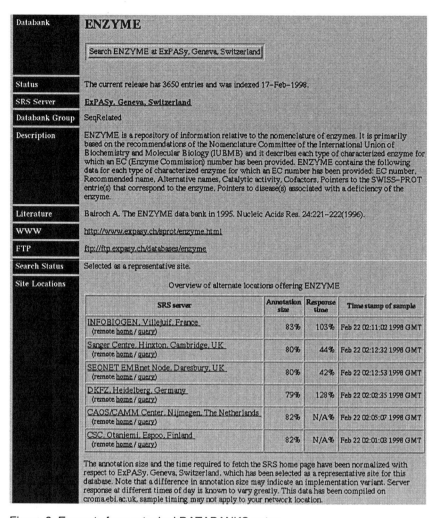

Databank	**ENZYME**
	Search ENZYME at ExPASy, Geneva, Switzerland
Status	The current release has 3650 entries and was indexed 17-Feb-1998.
SRS Server	ExPASy, Geneva, Switzerland
Databank Group	SeqRelated
Description	ENZYME is a repository of information relative to the nomenclature of enzymes. It is primarily based on the recommendations of the Nomenclature Committee of the International Union of Biochemistry and Molecular Biology (IUBMB) and it describes each type of characterized enzyme for which an EC (Enzyme Commission) number has been provided. ENZYME contains the following data for each type of characterized enzyme for which an EC number has been provided: EC number, Recommended name, Alternative names, Catalytic activity, Cofactors, Pointers to the SWISS-PROT entrie(s) that correspond to the enzyme, Pointers to disease(s) associated with a deficiency of the enzyme.
Literature	Bairoch A. The ENZYME data bank in 1995. Nucleic Acids Res. 24:221-222(1996).
WWW	http://www.expasy.ch/sprot/enzyme.html
FTP	ftp://ftp.expasy.ch/databases/enzyme
Search Status	Selected as a representative site.
Site Locations	Overview of alternate locations offering ENZYME

SRS server	Annotation size	Response time	Time stamp of sample
INFOBIOGEN, Villejuif, France (remote home / query)	83%	103%	Feb 22 02:11:02 1998 GMT
Sanger Centre, Hinxton, Cambridge, UK (remote home / query)	80%	44%	Feb 22 02:12:32 1998 GMT
SEQNET EMBnet Node, Daresbury, UK (remote home / query)	80%	42%	Feb 22 02:12:53 1998 GMT
DKFZ, Heidelberg, Germany (remote home / query)	79%	128%	Feb 22 02:02:35 1998 GMT
CAOS/CAMM Center, Nijmegen, The Netherlands (remote home / query)	82%	N/A%	Feb 22 02:05:07 1998 GMT
CSC, Otaniemi, Espoo, Finland (remote home / query)	82%	N/A%	Feb 22 02:01:03 1998 GMT

The annotation size and the time required to fetch the SRS home page have been normalized with respect to ExPASy, Geneva, Switzerland, which has been selected as a representative site for this database. Note that a difference in annotation size may indicate an implementation variant. Server response at different times of day is known to vary greatly. This data has been compiled on croma.ebi.ac.uk, sample timing may not apply to your network location.

Figure 9. Excerpts from a typical DATABANKS entry

Greater Ease of Administration

SRS server administrators create both parsers and documentation for databanks they offer on their servers. DATABANKS helps avoid duplication of effort by sharing parser and documentation files, and gives easy access to improvements by members of the SRS community.

Additionally, DATABANKS offers an easy way to check for database updates by comparing database sizes and release numbers between alternative sites. We also believe that DATABANKS may be of help to curators of centrally maintained database catalogues. They could utilize the extensive database documentation collected, which is assured to be current by automatic compilation of material from active servers.

SRS Servers World Wide

To date, the EMBL-EBI's list of public SRS servers contains 35 sites in 24 countries. Together they provide the scientific community with access to around 250 different databases and over 1000 databank copies. Each SRS server locally stores such a list of known public servers. When a local version of DATABANKS is compiled at a site, the lists from all visited servers are inspected and the site's local list is extended by including any new servers found. This removes the necessity of a central site to which new public servers must be reported. The association of public servers thus functions as a distributed system without the need for a centralized or hierarchical structure. To our knowledge, already 6 sites compile their local versions of DATABANKS. DATABANKS provides users with an up to date direct gateway into the ever growing network of databanks, whilst making life easier for server administrators.

Conclusion and Future Works

The SRS system is widely used in the bioinformatics and molecular biology community to access biological information in the form of flat files. We can ascribe this to a successful model of collaborative integration, where the SRS system administrators collaborate by exchanging structure descriptions and parsers for the databanks. This has also been possible thanks to the easy descriptions of flat file databanks using the Icarus language and the clear separation of site specific configurations from databank specific information.

In this chapter, after outlining the core features of SRS: parsing, indexing, linking and querying, we focused on recent developments concerning the World Wide Web interface, applications, views and the Databank of DATABANKS. However, there is more to come as the SRS system is continuously evolving. Current developments deal with providing permanent SRS sessions that a user or a group of users can create and then access later, and also generating SRS wrappers for different kinds of clients in a client/server architecture.

The client/server paradigm is based on the conceptual distinction between 'servers' that are stand alone software components that provide 'services' (data and operations that manipulate this data) and 'clients' that use these services. The first wrappers we provide are CORBA wrappers. CORBA (Common Object Request

Broker Architecture [18]) is an open standard considered a good solution for the development of Client/Server applications in distributed heterogeneous environments (different computers, different operating systems, different programming languages). Our CORBA wrappers allow client applications (e.g. visualization tools) to access SRS servers remotely (through the Internet) through an Object Request Broker (ORB). These wrappers are generated based on 'loaders' defined in Icarus by SRS administrators on the server side.

We might also consider a more flexible model of client-server, in which the client application can send to the server information about the granularity of requested information, and negotiate dynamically a communication protocol by sending a loader specification from the client side. This would be very useful in a field where client applications are developed by groups that have little interaction with the database maintainers. Finally, we are studying the integration of object oriented and relational database management systems. This will allow mirrored data from external sources to be kept in flat files, while local data are handled by a database system. The interfaces should allow work on both systems in the same way, with hybrid query resolution and query optimization routing the query to the more appropriate system.

References

1. Interpreter of Commands And RecUrsive Syntax
   http://srs.ebi.ac.uk:5000/man/srsman.html
2. Stoehr, P.J. and Cameron, G.N., *The EMBL Data Library,* Nucleic Acids Res., 1991, pp.2227-2230
3. Burks, C., Cassidy, M., Cinosky, M.J., Cumella, K.E., Gilna, P., Hayden, J.E.-D, Keen, G.M., Kelley, T.A., Kelly, M., Kristoffersson, D. and Ryals, J., *Genbank,* Nucleic Acids Res., 1991, pp. 2221-2225
4. Johnson, S., *YACC: Yet Another Compiler Compiler,* Supplementary Documents 1. University of California, Berkeley, 1986
5. Lesk and Schmidt, *A Lexical Analyzer Generator,* UNIX Programmer's Manual: Supplementary Documents 1. University of California, Berkeley,   1986
6. Wirth, N., *CompilerBau.* B.G. Teubner, Stuttgart, 1984, pp.34-44
7. Bairoch, A. and Boeckmann, B., *The SWISS-PROT protein sequence data bank,* Nucleic Acids Res., 1991, pp.2247-2249
8. Abola, E. E., Bernstein, F.C. and Koetzle, T.F., *The Protein Data Bank,* Computational Molecular Biology. Sources and Methods for Sequence Analysis., 1988, pp. 69-81
9. Etzold, T. et al, *SRSWWW version 5.1 at EMBL-EBI.* http://srs.ebi.ac.uk:5000/
10. Etzold, T., Ulyanov, A. and Argos, P., *SRS: Information Retrieval for Molecular Biology Databanks,* Methods In Enzymology, 1996, pp. 114-128
11. Etzold, T. and Verde, G., *Using Views for Retrieving Data from Extremely Heterogeneous Databanks,* Pac. Symp. Biocomput., 1997, pp. 134-141.

12. Altschul, S.F., Gish, W., Miller, W., Myers, E.W., and Lipman, D.J., *Basic Local Alignment Search Tool,* J. Mol. Biol., 1990, pp.403-410

13. Thompson, J.D., Higgins, D.G., and Gibson, T.J., *CLUSTALW: Improving the Sensitivity of Progressive Multiple Sequence Alignment through Sequence Weighting, Position-Specific Gap Penalties and Weight Matrix,* Nucleic Acids Res., 1994, pp. 4673-4680

14. A. Bairoch, *Nucleic Acids Res.* **24**, 1996, pp. 221

15. Kyte, J. and Doolittle, R.F., *J. Mol. Biol.,* 1982, pp. 105

16. Leipe, D. and Soussov, V.,
http://www3.ncbi.nlm.nih.gov/Taxonomy/taxonomyhome.html

17. Etzold, T. et al, *SRS version 5.1 ftp site*
ftp://ftp.ebi.ac.uk/pub/software/unix/srs/srs5.1.0.tar.gz

18. Object Management Group (OMG), *Common Object Request Broker: Architecture and Specification, Revision 2.0,* 1995

19 BIOLOGY WORKBENCH: A COMPUTING AND ANALYSIS ENVIRONMENT FOR THE BIOLOGICAL SCIENCES

Roger Unwin, James Fenton, Mark Whitsitt,
Curt Jamison[1], Mark Stupar, Eric Jakobsson
and Shankar Subramaniam[2].

National Center for Supercomputing Applications
University of Illinois, Urbana Champaign
Urbana, Illinois

Introduction

BioInformatics is playing an ever-increasing role in modern biological research. With the advent of large-scale high-throughput genome sequencing and the availability of complete genome sequences from several species (Akiyama and Kinehisa, 1995), accessing and processing of the data has become the largest bottleneck in biology. Databases like GenBank, PIR, and Swiss-prot provide valuable repositories for the vast quantities of sequence information, while computational tools like BLAST (Altshul et al., 1990), FASTA (Pearson and Lipman, 1988), and GenQuest (Shah et al., 1994) allow the researcher to access that information.

A large number of computational biology tools exist as resources freely available to biologists through the Internet and the World Wide Web (WWW). Other tools exist as readily available "shareware" which biologists can download and install on their own computers. Novel computational methods are constantly being developed and incorporated into new software.

The usefulness of these programs to the general biologist is limited by the ability of the researcher to input and transfer data between the databases and the

---

[1] Current Address: National Institutes of Health Bethesda, MD

[2] Corresponding author

computational tools. Most programs have been written to work with a particular database. Because the major biological databases have different formats, often the researcher is required to manipulate the output from one program before transferring it to another analysis program.

A common example of the problem represented by non-interoperable software is found in the analysis of a sequence from an automated DNA sequencer. The sequence is passed from a workstation, often a Macintosh computer, to a Unix workstation where it is preprocessed for high probability coding sequences and then filtered through BLAST to determine homology with any existing sequences. The latter computation is often done on a high-end computer. The entire process necessitates numerous FTP operations, file format changes and manual input of parameters to run the programs. It also requires a large investment of time by the Biologist in learning and carrying out routine computer operations.

Another difficulty facing researchers who wish to utilize new computational tools is the hardware requirement of the programs. While most tools are freely available, many are written for the UNIX operating system, due in large part to the computational power of UNIX workstations. The majority of biological researchers come from independent laboratories at small colleges and universities, and generally have neither the resources nor expertise to maintain large UNIX systems. Instead, most researchers rely upon personal computers which usually do not have the ability to perform the complex calculations required by many modern tools in a timely fashion.

There has been, therefore, a need for a seamless environment for biological data access and analysis. The Biology Workbench was developed to fulfill this need **(http://biology.ncsa.uiuc.edu).** The Biology Workbench uses CGI scripts and programs to integrate databases and tools into the web environment. Querying, analyzing and computing can, therefore, be done on a server from any client that is connected to the Internet and has web access.

The Workbench is object-centric. The objects range from high level objects such as session tools, databases and analysis programs to fine-grained objects such as protein and DNA sequences and three-dimensional structures. All necessary object manipulation tools are embedded in PERL and C programs, which pass the correctly formatted objects to the analysis programs.

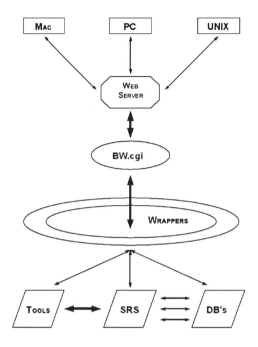

Figure 1.   Organization and Flow of Information through the Biology Workbench.

The WWW interface of the Biology Workbench provides a uniform user interface which most researchers are already familiar with, thus reducing the amount of computer knowledge needed to utilize the tools and perform database searches.

System and Methods

The Biology Workbench is a set of Common Gateway Interface (CGI) applications written in PERL (Wall and Schwartz, 1991) and C languages.

The Workbench uses a Netscape Web Server, which can be run on any small workstation, and a Compute Server, which is a multiprocessor high-end computing platform. The Workbench is accessible from any client which is web-browser compatible and is networked. Any Web Server, such as Netscape Navigator or Microsoft Internet Explorer can be used.

BWB and the HTML environment

The core of the Biology Workbench is the BW.cgi program, responsible for handling, among other tasks, the communication between the tools and the web server. Many functions of the Workbench are hard-coded into the core, but the majority of the functions relating to tool integration remain outside the core, allowing for integration and customization of programs and tools. Many of the external subroutines rely on core functions, but a full description of these core functions is beyond the scope of this paper.

The web-browser environment is a "stateless" system. That is, one page normally has no idea whatsoever what data the previous page presented, or what the next page will contain. This makes continuous analysis of a specific data set across a number of HTML pages difficult. The Workbench overcomes this limitation by storing information on the pages as "hidden" fields and by keeping sequence data in user specific session files, stored on the Compute Server.

Module Use

Use of most modules (or tools) follows the following simple pattern. The user selects a sequence or set of sequences, selects an analysis tool or module, and then clicks the "Perform Selected Operation" button. Next, the user is presented with a setup or parameters page. After parameter selections are made, the submit button is clicked and the results are returned. These results may consist merely of textual information, or may contain sequence data that can be viewed and imported to the Workbench. While this path is not always taken, it does describe how the majority of the tools are used.

Program Input and Output

Most of tools currently integrated into the Biology Workbench were originally stand-alone, command-line, textual input/output programs. Perl (Practical Extraction and Report Language), with its incredibly strong text handling features, is perfect for handling and parsing the textual I/O from the tool. The HTML code necessary can be easily generated to reformat the data for submission to the next tool.

Because the tools in the Workbench have been written by a variety of authors, the types and styles of input and output are equally variable. It is, therefore, vital to fully understand the command line parameters, input formats, output options (STDOUT, output file, or both), as well as the significance of the I/O to the user.

While sending textual output to the HTML page might be sufficient for some tools, other tools offer the opportunity for importing their results back to the Workbench, therefore requiring more involved HTML programming to allow the user to view and select sequences. The HTML environment also allows for enhanced output presentation that can aid the user's interpretation of their results.

Standard Subroutines

In order to make wrapper development easier and standardize tool integration into the Workbench, several files containing subroutines are included in the Workbench tree. These files are generally "required" by all Perl wrappers, but the subroutines may be extracted and revised to suit individual needs. While we recommend using the standard subroutines wherever possible, we cannot fully predict the needs of users for other non-standard tool specific routines. This concept is being refined for future releases of the Workbench.

The central file for tool integration is the "html.pl" file, which contains the necessary subroutines for communication with BW.cgi (the Workbench core CGI program) and for passing configuration information to the wrapper scripts. This file should be included at the beginning of all scripts using the "require" directive in Perl. Tools wrappers may, however, require only a subset of the subroutines in the html.pl file.

Sessions

Each time a user accesses the Workbench, a previous analysis session can be resumed or a new one initiated. The role of the sessions is to maintain the state and sequence of operations performed on the Workbench during an interactive session. Sessions can be useful for both recalling previously performed operations as well as for comparison with newer analyses. Essentially, sessions provide the history of the Workbench use and can be updated, erased or permanently stored.

Databases

Almost all of the relevant biological sequence, structure and literature databases are accessible from the Biology Workbench. The databases are federated by identification of common objects and indexing each database based on these objects. We use the SRS schema (Etzold and Argos, 1993) to store and access the databases allowing them to be queried for multiple objects as well as combinations of objects using Boolean operators on the query construction page. The Workbench also permits nested Boolean queries.

To aid the integration of SRS indexed databases, the file "db_config.pl" containing routines for parsing database directories and identifying available databases is provided. These routines also extract important information on specific databases to allow sequence information to be properly imported to the Workbench.

A database mirroring program has also been developed for the Biology Workbench, which fetches databases as they are updated in their primary repositories. The indexing and creation of Blastable sequence database files is done locally each time new databases are retrieved. A complete list of all databases available on the Workbench is presented in Appendix A.

Analysis Modules

238

The Biology Workbench contains all the commonly used, publicly available analysis modules for the protein and nucleic acid sequences and structures. "Bread-and-butter" modules for biologists such as BLAST and FASTA, programs for aligning multiple sequences such as CLUSTALW and MSA and multiple sequence profile and statistical analysis tools are all a point-and-click away on the Workbench. A complete list of available tools is presented in Appendix B.

System Overview

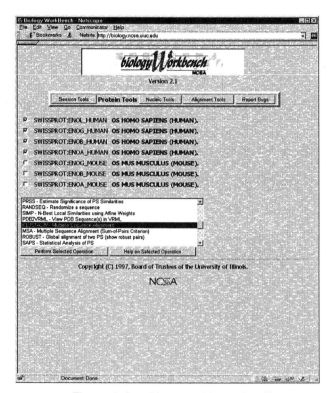

Figure 2: Invoking an old session file

A typical user session begins with a user-specific login to the Workbench server followed by the invocation of a session. Invoking an old session file places the user in the final state of the most recent prior invocation of that session. The user may then choose one of the analysis tool-sets, namely Protein Tools, Nucleic Acids Tools, and Alignment Tools. Selection of the Protein Tools, for instance, displays a page in the browser window containing a menu of protein-based tools, and a list of previously imported protein sequences (Vide Fig. 2.).

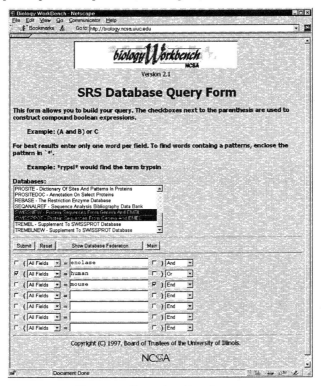

Figure 3: Querying Databases

The user may access existing databases by selecting the "SRS Multiple Database Search" menu item. A page containing the menu of databases pertinent to the current tool-set (Protein or Nucleic Acids) is displayed (Vide Fig. 3). The user may then construct a database query for any pertinent object or combination of objects, submit the query to the server, and the resulting page returned to the browser will contain the results of the query. The user may then import a selected set of sequences from the search results to the Workbench for further analysis. For instance, the user may choose a subset of previously imported "enolase" protein sequences from a particular species and invoke the program CLUSTALW to generate a multiple sequence alignment. Invoking the program with a click, displays a web page containing the standard program parameters which can be altered by the user. The sequences are then submitted for computation using the selected parameters. As soon as the alignment computation is complete, the results are displayed in the browser window. In this instance, a set of aligned enolase sequences along with a phylogenetic tree would be presented. The aligned sequence set can then be imported back into the

240

Workbench for color-coding based on degree of similarity using a program such as MSA-Shade. Once the color mapping is completed, a GIF image of the color-coded sequence alignment is displayed in the browser page. Alternatively, a postscript file of the color coded alignment may be downloaded for publication purposes (Vide Fig. 4).

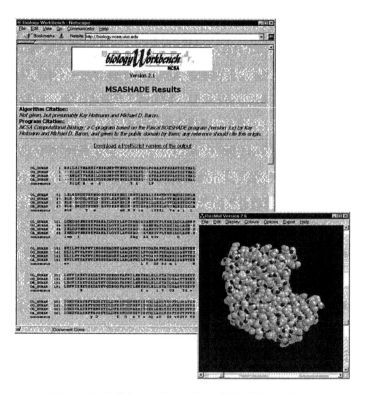

Figure 4: MSA results and protein 3D structure

The user may also input their own sequences in several formats, either manually or as an uploaded file, for analysis on the Workbench. Many of the computing tasks in the Workbench may also be done as a batch submission on the Compute Server by selecting the "batch" option. These results will then be made available to the user either during the current Workbench session or upon resuming the session at a later time.

Much of the display on the Workbench is tailored to the program outputs of the specific tools. For instance, using an appropriate MIME-type for RASMOL, three dimensional structures of macromolecules can be displayed on a browser window(Vide Fig. 4).

Discussion

The Biology Workbench is a WWW-based analysis environment which brings high-powered computation into an experimental laboratory. The seamless integration of remote database and computational biology tools forms creates a virtual computer, with WWW browsers as the GUI interface. The platform independent access to tools creates an environment in which all biologists can utilize programs which previously were found only in large laboratories with heavy bioinformatics investments.

The future of the Biology Workbench is targeted towards genomics. Several aspects of genome informatics including automated analysis and annotation, newer tools for protein and gene clustering, and tools for fast comparisons across genomes are all under development.

Acknowledgements

This work was supported in part by NSF grant ASC 89-02829 and by grants from MDL Informatics Corporation. All the computational resources for the public version of the Biology Workbench are provided by NCSA.

APPENDIX A

List of All Currently Available Biological Databases in The Biology Workbench

Nucleic Acid Databases
GenBank - NCBI Nucleic Acid Sequence Database (Protein also)
GenBankNew - NCBI Nucleic Acid Sequence Database (Protein also)

Protein Databases
BLOCKS - Aligned Ungapped Highly Conserved Protein Segments
NRL3D - Sequence and Annotation Information Taken From The PDB
PDBFINDER - Protein Sequences From The PDB
PIR - The Protein Information Resource
SWISSPROT - Protein Sequences From Geneva And EMBL
SWISSNEW - Protein Sequences From Geneva And EMBL

TREMBL - Supplement To SWISSPROT Database
TREMBLNEW - Supplement To SWISSPROT Database
PIRALN - Alignment Database from The Protein Information Resource

Literature and Textual Databases
ECDC - Information On The E.coli K12 Chromosome
ENZYME - Data For Enzymes With An EC Numbers
EPD - Eukaryotic Promotor Database
FlyGene - Subset Of FlyBase
PRODOM - Comprehensive Collection Of Protein Families
PROSITE - Dictionary Of Sites And Patterns In Proteins
PROSITEDOC - Annotation On Selected Proteins From PROSITE
OMIM - The Mendelian Inheritance In Man Data Bank
REBASE - The Restriction Enzyme Database
SEQANALREF - Sequence Analysis Bibliography Data Bank

APPENDIX B

List of Currently Available Tools in the Biology Workbench

Sequence Management Tools
Add New Sequence
Edit Sequence
View Sequence
Delete Sequence
Copy Sequence
Download Sequence
View Scoring Matrices

Common Analysis Tools

SRS	Multiple database searches with Boolean queries
FASTA	Heuristic database similarity searching
SSEARCH	Smith-Waterman database searching
ALIGN	Smith-Waterman pairwise global sequence alignment
LALIGN	Smith-Waterman pairwise local sequence alignment
LFASTA	Heuristic pairwise local sequence alignment
BESTSCOR	Self-comparison similarity scores
RANDSEQ	Sequence randomization for statistical analysis
CLUSTALW	Multiple sequence alignment and phylogenetic tree construction

PROFILESCAN   Comparison of protein or DNA sequences with profile libraries

MOTIFGREP   Sequence identification using Regular Expressions

MOTIFGREPDB   Database searching using Regular Expressions

Protein Tools

BLASTP   Heuristic database similarity searching

TBLASTN   Heuristic database similarity searching (protein query, database translated)

TFASTA   Heuristic database similarity searching (protein query, database translated)

TFASTX   Heuristic database similarity searching (protein query, database translated)

PRSS   Protein sequence similarity statistical significance calculation

CHOFAS   Protein secondary structure prediction - Chou and Fasman

GOR4   Protein secondary structure prediction - Garnier, Osguthorpe, and Robson

GREASE   Kyte - Doolittle hydropathicity profiles

ROBUST   Global pairwise sequence alignment

SAPS   Statistical analysis of protein sequence composition

SIMP   Smith-Waterman pairwise sequence alignment

AASTATS   Amino acid composition statistics

MSA   Multiple sequence alignment

PROSITE   Searching PROSITE database for protein sequence patterns

PROSEARCH   Searching PROSITE database for protein sequence patterns

MPSSP   Combined algorithm protein secondary structure prediction

HTH   Helix-Turn-Helix identification

DSSP   Secondary structure and solvent accessibility predictions

Nucleic Acids Tools

BLASTX   Heuristic database similarity searching (translated nucleic acid query, protein database)

BLASTN   Heuristic database similarity searching (nucleic acid query, nucleic acid database)

TBLASTX   Heuristic database similarity searching (translated nucleic acid query, translated nucleic acid database)

FASTX   Heuristic database similarity searching (translated nucleic acid query, protein database)

REVCOMP   Generate the reverse-complement of a nucleic acid sequence

SIXFRAME   Translate a nucleic acid sequence in all six reading frames

TACG   Restriction enzyme recognition site identification

TACGXLATE   Translate a nucleic acid in a user defined reading frame

Alignment Tools

MSASHADE	Formatting and color-coding multiple sequence alignments
TMAP	Identification of transmembrane protein segments
CLUSTAL_W	Phylogenetic tree analyses
PROTDIST	Evolutionary distance matrix computations
PROTPARS	Unrooted phylogeny inference

References

2. Akiyama, Y. and Kinehisa, M. (1995) Introduction to database services on the GenomeNet. Exp. Med., 13 (in Japanese). See also URL http://www.genome.ad.jp/dbget/dbget.links.html

3. Altschul, S.F., Gish, W., Miller, W., Myers, E.W., and Lipman, D.J. (1990). Basic Local Alignment Search Tool. J. Mol. Biol. 215, 403-410.

4. Etzold, T., and Argos, P. (1993) SRS - an indexing and retrieval tool for flat file data libraries. CABIOS 9, 49-57.

5. Pearson, W.R., and Lipman, D.J. (1988). Improved tools for biological sequence comparison. Proc. Natl. Acad. Sci. USA 85, 2444-2448.

6. Shah, M.B., Guan, X., Einstein, J.R., Matis, S., Xu, Y., Mural, R.J., and Uberbacher, E.C. (1994). User's guide to GRAIL and GENEQUEST. URL

7. http://avalon.epm.ornl.gov:80/manuals/grail-genequest.9407.html

8. Wall, L., and Schwartz, R.L. (1991). Programming perl. O'Reilly & Associates, Sebastapol, CA, USA.

20 EBI: CORBA AND THE EBI DATABASES

Kim Jungfer, Graham Cameron and Tomas Flores

EMBL Outstation - Hinxton, The European Bioinformatics Institute, Wellcome Trust Genome Campus, Hinxton, Cambridge CB10 1SD, United Kingdom

Introduction

The European Bioinformatics Institute (EBI) is a major center for biological data. Research groups have collected genome-related data for the last 15 years, during which the amount of data has grown exponentially. There are now more than 300 publicly available collections of highly interrelated data. The more the size and complexity of molecular biology data grow, the more important become automatic tools for management, querying and analysis. The current limitations in using this wealth of information are not due to missing technology but to lack of standardization. Biologists utilize every possible hardware platform, operating system, database management system and programming language. The de facto standard CORBA [9] [10] [11] offers the opportunity to make such differences transparent and thereby helps to combine disparate data sources and application programs.

Molecular biology data have traditionally been stored in simple text files often referred to as flat-files. Flat-files are the minimalist storage mechanism, adopted for small data sets and simple programs – easily distributed and readily comprehensible. Even large volumes of complex data, although managed in database management systems, are often distributed as flat-file "entries". This leads biologists to see the flat-file as the basic data representation. The advent of the World Wide Web [2] strengthened this view. Flat files can easily be transformed to hypertext by turning references into hypertext links. The Sequence Retrieval System SRS [7] is a well-known example of this approach. Flat-files became the center of the data flow in molecular biology. Every data collection has to provide a flat-file version in order to distribute the data and most analysis programs use flat-files as their data source.

This central role of flat-files has several disadvantages. The most obvious problem is that flat-files are difficult to use. Writing a parser is a non-trivial task, which is further complicated by imprecisely specified and frequently changing formats. Previous attempts to establish a standard flat-file format have failed because biologists do not agree on a single model of their data. Another major draw back is that flat-files can lead to an immense waste of computing resources. Different programs often expect different flat-file formats so that the same site needs to keep multiple copies of the same data (for example FASTA and BLAST). Finally, the wish of human readability results in the attempt to keep all information associated with an entry together. Since this attempt conflicts with the goal of normalization, flat-files tend to be redundant. For instance the classification of a virus might be repeated in all entries from that virus.

Strategy

It would not be realistic to abolish the usage of flat-files in general, but it is necessary to allow for alternative solutions to overcome the problems discussed above. This can be achieved only if the dependency on flat-files and their formats is removed. The task for which flat-files are least suited - providing a programming interface to application programs - demonstrates this point. A sequence comparison program, which depends on a flat-file format, violates the principle of data independence (see for example [6]). In an ideal world it should make no difference whether an application accesses a local flat-file or a remote relational database as long as both serve the same data. The concept of interfaces helps to achieve this goal. Data source and application programs interact through interfaces, which hide the underlying implementation details. A piece of software, which can be used in different contexts through a defined interface, is called a component. But componentry can work only if used together with a widely accepted standard for defining interfaces and invoking methods through them. CORBA is such a standard.

CORBA

In 1989, the Object Management Group (OMG) was formed [19]. It now has more than 700 members, including practically all major software vendors, hardware vendors and large end-users. OMG's stated goal is to standardize and promote object technology. The core specification adopted by the OMG is the Common Request Broker Architecture – CORBA; references [12] [14] give good introductions. CORBA combines the concept of interfaces with the object oriented distributed programming paradigm. Among other things it specifies:

- the Interface Definition Language (IDL), which provides a language-independent way of describing the public interface of objects.

- the Object Request Broker (ORB), which transparently transmits request from clients to object implementations.

Once the interface for a distributed object is defined in IDL, any client can use it depending only on the objects public interface. It is irrelevant to the client where the object implementation resides or in what language it is implemented. In order to invoke methods on the CORBA object, the client needs to obtain an object reference for that object. The object reference serves as a proxy object on which the client can invoke methods as defined in IDL

It is worth comparing the CORBA approach with the more familiar Remote Procedure Call (RPC). Using RPC a client invokes a function on a server, while in CORBA a client invokes a method on an object. Such a CORBA object is a purely logical entity whose implementation is unknown to the client. In many cases the server implements several CORBA objects as shown in Figure 1.

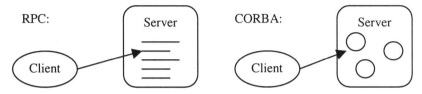

Figure 1: RPC versus CORBA

Microsoft's Distributed Component Object Model (DCOM) [3] is an important alternative to CORBA. However, DCOM is not considered here because of its proprietary nature and lack of cross-platform support.

Interfacing Databases with IDL

How should interfaces to databases look? Should the data model be reflected in IDL, and if yes how? Since the answer to these questions depends very much on the concrete application, the best way to find the "right" IDL is very often a use-case analysis [8]. Several important cases will be examined in this section using the simple example of protein sequences. It is important to understand that these different possibilities do not exclude each other – they simply meet different requirements of different applications. In our example, every protein sequence has a number of attributes, one of which is a reference to an entry in the EMBL Nucleotide Sequence Database.

"Interface"-Based

```
Interface SwissProtSequence {
    attribute string accession;
    attribute string description;
    attribute string sequence;
    attribute EMBLSequence EMBL;
}


Interface EMBLSequence { … }
```

This is perhaps the most natural representation. Every class of the data model is represented by a corresponding interface. Every database entry is represented as a CORBA object. The application program can use such a sequence object as if it were local. Access to the attribute "EMBLSequence" will transparently cross database boundaries. Interface-Based data representations are therefore very easy to use. The biggest disadvantage is that this approach might lead to heavy network traffic. If the application program wants to get all four attribute-values then four remote method invocations will occur. If the application requires access to many such objects the load becomes unacceptable. Since the data are accessed in small pieces, this approach is often called "fine-grained".

"Struct"- Based

```
struct ProteinSequence {
    string accession;
    string description;
    string sequence;
    string EMBLAccesion;
}


struct EMBLSequence { … }
```

Every database entry is represented by a "struct". In contrast to CORBA objects, structs are passed by value. Access to the individual fields is therefore local and very fast. On the other hand the application has to get the whole struct even if it is only interested in the sequence field. Structs are nearly as convenient to use as CORBA objects with one subtle difference. Structs can't be referenced physically. In this example the SwissProt entry contains the accession number of an EMBL entry. The client has to know itself how to get the EMBL sequence given the accession number. The upcoming CORBA 3.0 specification will address the pass-by-value issue and make the definition of structs unnecessary in some cases.

Generic

In this approach the data model is not represented in IDL. The objects are instead encoded in generic types like strings. The main advantage of this approach is ease of maintenance. Using IDL, clients and servers have to be compiled together with the classes, which are generated from the IDL definition. This can become a burden if the data model is large or changes frequently. Generic representations are well suited for generic applications, which are not interested in the semantics of the data. For end-user applications (e.g. an alignment program) generic representations are very cumbersome to use. An example of a syntax of this sort is the object interchange format OIF, defined in the ODMG 2.0 standard [4].

Queries

The ability to express queries is important for all databases. This aspect is standardized in CORBA through the query service specification [10]. The central element is a CORBA object of the type query evaluator. A query evaluator takes a query string as input and returns a reference to a result collection. This collection is another CORBA object, which can itself implement the query evaluator interface. Because the query service does not specify the elements of the result collection, it can be combined with any of the above listed representations. Furthermore the query service can be used with every possible query language depending on the underlying database management system.

Conceptual Data Model

Even though Interface-Based and Struct-Based representations can have a very strong similarity to the conceptual data model, it is important not to confuse them. The purpose of the conceptual model is to model the data as they really are, using concepts close to humans. Neither the database schema nor the IDL can replace such a high level model. The database schema is specific to the used DBMS and may contain optimizations. The IDL is independent from implementation details but it represents merely an application specific view.

250

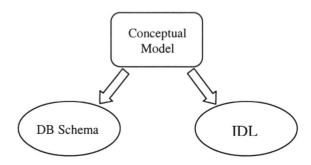

Figure 2: Relationship of Conceptual Model, DB Schema and IDL

CORBA Wrappers

Generally, today's databases do not provide CORBA interfaces to their data. It is therefore necessary to build a so-called wrapper – a program which implements IDL interfaces for already existing "legacy systems". This section will discuss some of the technical issues involved for a fine-grained CORBA interface.

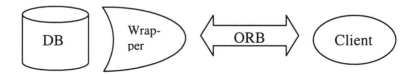

Figure 3: CORBA wrappers make existing DBs available through the ORB

Registering Large Numbers of Objects

Databases can contain millions of entries. The ORB has to be able to keep track of the connection between these database entries and the corresponding CORBA object references. The general idea is to encode a unique identifier of the database entry in the object reference. It is therefore essential that such an object identifier or key attribute exists in the database. Most ORBs have their own proprietary way of implicitly registering large numbers of objects. The recently adopted Portable Object Adapter (POA) standardizes this possibility.

Wrappers with and without State
Typically every CORBA object is represented by one corresponding object of the implementation language of the server (e.g. a C++ object). In the case of a database wrapper this means that the C++ server contains a copy of the database entry (Figure 4a). If a client invokes for the first time a method on an object reference the corresponding database entry is loaded and the C++ object is instantiated. It stays in memory for subsequent method invocations until the server actively removes it to avoid maintaining a copy of the whole database. This approach can be fast because the CORBA server is a cache for database entries but it makes a system for garbage collection necessary.

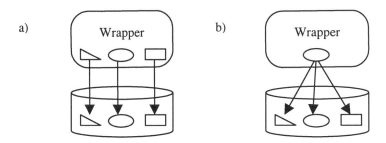

Figure 4: CORBA wrappers with a) and without state b)

In the second possibility (Figure 4b) the CORBA server is stateless. One language object implements a set of different CORBA objects each representing another database entry. For every request the server extracts the object identifier from the request in order to find out which CORBA object it has to emulate. In this approach no garbage collection is needed but every request requires access to the underlying database.

Object-Relational Mapping
Relational database management systems (RDBMS) are still the standard today. Several major databases at EBI, such as the EMBL Nucleotide Sequence Database and TREMBL, use the RDBMS ORACLE. Since the relational model is very different from an object-oriented conceptual model, object-relational mapping [15] becomes the central problem for CORBA wrappers. There are several tools, which can help to define object-oriented views on top of a relational database. An example is the object-relational mapping tool "Persistence" [20]. It follows the philosophy of object-oriented databases [1] [4] in the sense that it makes database entries directly available as C++ objects. Persistence is therefore well suited for implementing CORBA servers, which cache database entries in main memory. The main disadvantage of this tool is that the developer has not much freedom to define object schemas. The Persistence object schema is very close to the original table structure of the relational database. Relational views and a certain amount of hand coding are often necessary to create the required mapping. Another type of mapping tool supports the mapping of object-oriented queries to SQL queries. (for example OPM

[5]). Such a tool is potentially better suited for stateless wrappers and for the implementation of a CORBA Query Service.

CORBA Prototypes

There are now a large number of CORBA applications and prototypes available at EBI [17]. They include:

Database Wrappers

- **EMCORBA** A fine-grained CORBA interface to the EMBL Nucleotide Sequence Database by Jeroen Coppieters, Carsten Helgesen, Philip Lijnzaad and Timothy Slidel. The server was implemented in C++ using the RDBMS ORACLE, the object-relational mapping tool Persistence and the ORB ORBIX [18].

- **RHdb** There are two CORBA servers to the Radiation Hybrid Database by Carsten Helgesen, Philip Lijnzaad and Patricia Rodriguez-Tomé. One is implemented in C++ using ORACLE, Persistence and ORBIX. The second server is implemented in Java using JDBC to query the ORACLE database and the ORB VisiBroker for Java [21]. The object-relational mapping was purely hand-coded.

Applications

- **AppLab** A general-purpose CORBA wrapper for a large class of command-line-driven applications by Martin Senger. It uses the GCG configuration file language.

- **Mapplet** A Java Applet, which displays radiation hybrid maps by Kim Jungfer. The CORBA server for this applet uses the flat-file version of RHdb.

- **Hyperbolic Viewer** A Java applet for the visualization of tree like data structures by Alan Robinson. In this example a taxonomy tree is displayed. Chris Dodge and Timothy Slidel provided the CORBA taxonomy server using Sun's ORB NEO and the RDBMS Sybase.

- **Genome Builder** A tool for building virtual sequences using the EMCORBA server by Juha Muilu.

- **Protein Sequence Space Viewer** Displays families of proteins and their residues in a multidimensional space, projected into 2D and 3D by Chris Dodge. The algorithm is computationally intensive and therefore separated from the Java viewer in a specialized CORBA server.

Discussion

CORBA and IDL minimize dependencies between different components by hiding implementation details like operating system, network, location and programming language. This decoupling of clients and servers dramatically facilitates the development and maintenance of application programs. Even though this is obviously an improvement of the current state of affairs, it pushes the problem to another level. How can software components developed by independent groups interoperate? They can do so only if they share a set of common IDL definitions.

OMG's Domain Task Forces (DTF) provide formal frameworks for the adoption of standard interfaces for a specific application domain. At the OMG meeting in Dublin, in September 1997, a "Life Sciences Research" Domain Special Interest Group (DSIG) was formed. The DSIG represents a first step to the creation of a DTF. Its goal is to provide a forum for everybody who wants to get involved in the creation of standards for the life sciences area [16]. Life Sciences Research includes, but is not limited to, such fields as genomics, bioinformatics and computational chemistry.

The standardization of interfaces will bring us eventually closer to the vision of componentry [13] where the functionality of today's monolithic, closed systems is taken over by collections of object services. We will be able to mix and match components from different providers and find uses for applications that were never dreamed of by their developers. This intrinsic openness will make it significantly easier for a lone researcher to contribute to the bioinformatics projects of the future.

Acknowledgements

We wish to thank Jeroen Coppieters, Chris Dodge, Carsten Helgesen, Katarzyna Kruszewska, Philip Lijnzaad, Phil McNeil, Juha Muilu, Jeremy Parsons, Nicole Redaschi, Alan Robinson, Patricia Rodriguez-Tomé, Martin Senger, Timothy Slidel and Anastassia Spiridou for their involvement in various CORBA projects at EBI.

The research into common CORBA interfaces on distributed data sources is supported by grant BIO4-CT96-0346 from the European Union (DG XII). The development of interfaces to distributed applications is supported by the BioStandards project, a consortium of EU (DG III), EMBL and 19 pharmaceutical companies.

References

1. Atkinson, M., Bancilhon, F., DeWitt, D., Dittrich, K. Maier, D., Zdonik, S. *The Object-Oriented Database System Manifesto*, First International Conference on Deductive and Object-Oriented Databases, 1989.
2. Berners-Lee, T., Cailliau, R., Luotonen, A., Nielsen, H.F., and Secret, A. *The World-Wide Web*, Commun. ACM 37(8), 1994, pp. 76-82.
3. Brockschmidt, K. *Inside OLE*, Second Edition, Microsoft Press, 1995.

4. Cattell, R. G. G., et al, *The Object Database Standard: ODMG 2.0,* Morgan Kaufmann, 1997.

5. Chen, I. A., and Markowitz, V. M. *An Overview of the Object-Protocol Model (OPM) and OPM Data Management Tools*, Information Systems, Vol. 20, No 5, 1995, pp. 393-418.

6. Date, C. J. *An Introduction to Database Systems*, Fifths Edition, Addison-Wesley, 1990.

7. Etzold, T., Ulyanov, A., and Argos, P. *SRS: Information Retrieval System for Molecular Biology Data Banks*, Methods in Enzymology v. 266, 1996, p. 144.

8. Jacobson, I., Christerson, M., Jonsson, P., and Overgaard, G. *Object-Oriented Software Engineering*, Addison-Wesley, 1992.

9. Object Management Group, *CORBA: Architecture and Specification*, OMG publication, 1996.

10. Object Management Group, *CORBAservices*, OMG publication, 1996.

11. Object Management Group, *CORBAfacilities*, OMG publication, 1996.

12. Orfali, R., Harkey, D., Edwards, J. *Instant CORBA*, John Wiley & Sons, 1997.

13. Sessions, R. *Object Persistence, Beyond Object-Oriented Databases*, Prentice Hall, 1996.

14. Siegel, J., *CORBA Fundamentals and Programming*, New York, John Wiley & Sons, 1996.

15. Wiederhold, G. *Views, Objects and Databases,* IEEE Computer, 19, 2, 1986.

16. http://lsr.ebi.ac.uk/

17. http://www.ebi.ac.uk/~corba/

18. http://www.iona.com/

19. http://www.omg.org/

20. http://www.persistence.com/

21. http://www.visigenic.com/

21 BIOWIDGETS: REUSABLE VISUALIZATION COMPONENTS FOR BIOINFORMATICS

Jonathan Crabtree, Steve Fischer, Mark Gibson and G. Christian Overton

Center for Bioinformatics, University of Pennsylvania, Philadelphia, PA 19104

Introduction

Genome centers around the world are ramping up to the sequencing throughput required to bring the first phase of the Human Genome Project to a close. It is more apparent than ever that bioinformatics support must play an integral role in genome-scale mapping and sequencing projects. As genomics data is collated into information management systems, it furthermore must become available to scientists in a form that facilitates their comprehension of it. Interactive data visualization components provide a way to fulfill this need

In large part because of the limited resources often allocated to bioinformatics in the past, those responsible for designing and implementing laboratory information systems have had to make do with what they could get, particularly in terms of obtaining software. As Goodman put it, "the choices that face the architect of a genome information system today are: (i) build it yourself so that it does exactly what you want, or (ii) adopt someone else's system and live with most of its quirks and limitations" [1].

The solution to the problem of limited resources is simple in principle: design systems as collections of interoperable components and share those components freely. The tacit assumption in this approach is that smaller, more modular programs will be able to avoid the quirks and limitations that tend to be the hallmarks of large "monolithic" systems. And even if some components fall short in this regard, making use of them is, by definition, not an all-or-nothing proposition. This is the motivation behind the bioWidgets project, an effort to create and share graphical user interface

(GUI) components for the display and visualization of genomic data, potentially over the World Wide Web.

Background: bioTk

A similar observation about the utility of component-based software led David Searls to create bioTk [2], a set of graphical user interface and data visualization tools. In bioTk, the direct progenitor of bioWidgets, Searls used the language Tcl/Tk to define visualization components (widgets, in Tk parlance) to display sequence, mapping and chromosome data. He provided a simple specification for invoking the widgets, as well as such ancillary components as menus, a context-sensitive help system, and utility dialog boxes. Searls also set a high standard with the documentation and tutorials that accompanied his toolkit. Figure 1 depicts the bioTk chromosome widget used in a simulated karyotype display.

New technologies: Java™

In the short time since Searls' work in Tcl/Tk, a number of significant new technologies have been introduced, among them Java™[3]. Typifying the growing popularity of the object-oriented paradigm in software engineering, Java™ is a language designed to provide a secure way to "program the web." Java™ programs called "applets" can be sent across the network along with web pages and a web browser will interpret the applets by running them on the user's local machine.

In the past, web-based interfaces to the databases and tools used in bioinformatics were greatly restricted in the interactivity they provided. A user might be able to request information and pose queries through a "fill in the blanks" form-based interface (e.g. retrieve the sequence with accession number x, or perform a BLAST[4] search against database y with sequence z). The information would be returned either as another web page (perhaps containing some "clickable" but otherwise-static images) or as an e-mail message. "Interacting" with the resulting data consists of navigating through it in a web browser or scrolling through the tens or hundreds of pages of results in a text editor.

Using Java™, however, it is possible to implement widgets like those found in bioTk and have them run—as applets—on any PC equipped with a web browser. Since Java™ support is an integral part of all major web browsers, it is possible to implement almost any imaginable user interface in this manner. For this reason, among others, the bioWidgets project has chosen an object-oriented approach to visualization component design and has standardized on the Java™ language for its specification and implementation efforts.

Visualization Solutions

Turnkey solutions have enjoyed ready acceptance and widespread use in the genomics community. ACEDB[5], for example, is a genomics database system which also provides an extensive visualization-oriented user interface. But, as noted earlier, using such systems entails an all-or-nothing commitment. Were they instead implemented in a component-based framework, they would be able to take advantage of new components developed by others. Furthermore, other developers would be able to extract useful components from these packages and reuse them in ways perhaps not originally envisioned by their designers.

There are a number of groups and individuals working on Java™-based graphical user interfaces for bioinformatics applications: Jean Thierry-Mieg's Java™ port of ACEDB, JADE[6]; the genome browser of Gregg Helt[7]; Andrei Grigoriev's DerBrowser[8]; and GDB's MapViewer[9] application are but a few examples.

What differentiates bioWidgets from these standalone applications is the intent to provide not just finished applications but also the underlying components and, more importantly, a framework or architecture for integrating them. This bioWidget architecture specifies how bioWidgets written by different developers interact to fit into the "big picture" of a finished application. It also empowers application designers to make choices at all stages of the design and implementation process. The decision to use a particular widget or set of graphical front-end components should not lock a developer into using a specific database system in the application's back-end.

To date, we have concentrated on developing displays for "linear" genomic data. In particular, we have constructed bioWidgets and applications for displaying the following kinds of data:

- Chromosomes and karyotypes (see Figure 1)

- Maps (physical, genetic, STS content, radiation hybrid, etc.)

- BLAST results (see Figure 2)

- Sequences (DNA and protein) and annotation (see Figure 3)

- Multiple map alignments (see figure 4)

- Sequence alignments

- DNA traces (i.e. ABI machine output)

The bioWidget Architecture

The bioWidget architecture is an adaptation of the Model-View-Controller (MVC) paradigm [10]. A "model" is an instance of domain data (e.g. DNA sequence data); a "view" is a visual representation of the model (an application may include more than one view of a model); and a "controller" interprets external input (e.g. from a user) and updates a model accordingly. When the model is updated the view receives notification and updates its representation accordingly.

In our case, the model is defined as a set of Java™ interfaces. An interface is an abstract specification of functionality in the form of a set of method signatures. Using interfaces allows us to define a model without detailing how the model will be implemented. A critical step in defining a widget is defining the model of which it will be a view. Because a widget is designed for reuse throughout the genomics community, its model must be generic enough so that it will not be inconsistent with data models used by would-be consumers. The philosophy therefore in defining models for bioWidgets is to keep them as simple as possible.

Here is the interface that defines the model of (ungapped) sequence used by our sequence widget:

```java
public interface SequenceReadOnly {
    public int getFirstChar();
    public int getLastChar();
    public String getSubSeq(int first, int last);
    public Interval[] getSelectedIntervals();
}
```

In this case, sequence is viewed simply as a string that can easily provide substrings and that also has a set of selected sub-intervals. Because this interface is easy to satisfy, it allows data to be read into the bioWidgets from flat files, databases (possibly using the JDBC™[11] database access specification), or even directly from other widgets or applications, using remote communication protocols such as RMI or CORBA[12]. Components called transducers do the job of translating data in disparate formats into objects satisfying the interface specifications recognized by the widget. Note that, in terms of performance, the above interface does not require that the sequence be *stored* as a string. It could be stored in a compressed format and decompressed on the fly in response to getSubSeq requests. It is the widget's responsibility to ask for only that part of the sequence it needs, so that the entire sequence need not be duplicated in every view.

In addition, the MVC approach ensures that any updates to the underlying databases or data sources are immediately propagated to the widgets' displays. Another feature of the bioWidget architecture—the use of multiple concurrent Threads of execution—ensures that the widgets will not freeze while performing such updates; the user will be able to continue to interact with the widgets while they incorporate new or updated data into their displays.

The architecture also exploits Sun's own generic component architecture, JavaBeans™[13]. JavaBeans™ are components written in Java™ that publish the information required by application builder tools. These builder tools allow users to interactively configure and connect beans. By requiring that the bioWidget models, views and controllers conform to the JavaBeans™ specification, we ensure that these components can be integrated into useful applications using standard JavaBeans™ integration tools.

bioWidget-Based Displays for Genome Annotation

Figure 3 depicts the graphical interface of GAIA[14], a system developed at CBIL to perform automated annotation of genomic sequence. Sequence and map bioWidgets are used in the system's web-based interface, which allows users to examine the annotation on sequences they have submitted. We will use GAIA to illustrate a number of the features provided by bioWidgets, particularly those that require communication between the bioWidgets and a surrounding application; i.e. those that depend on the bioWidget architecture.

The two widgets in the figure are displaying the same sequence entry from the GAIA database, the map widget (upper right) giving a high-level overview of the annotation, and the sequence widget (lower right) showing the actual DNA sequence, along with the same annotation in a slightly different form. From the perspective of the bioWidget architecture, the sequence widget is a view of a sequence model; and, both the sequence widget and the map widget are views of a shared model of sequence annotation. The widgets interact with the rest of the application (including the GAIA back-end and the web browser, which can be used to retrieve textual descriptions of the various analysis results displayed in the widgets) to provide a number of useful functions.

Coordinated Selection

Clicking on a particular annotation in the map widget selects (highlights) the annotation, and displays a short description in the widget's status bar. The sequence widget responds by jumping to the region of the sequence occupied by the feature and highlighting the corresponding DNA fragment. This behavior is a direct result of the sharing of an annotation model. In response to the user's actions, a controller marks that particular annotation in the model as "selected", and both views respond by highlighting and jumping to the annotation. The messages that travel between the model and the view to accomplish this are defined as part of the model's JavaBean™ specification.

Coordinated Scrolling and Zooming

The map viewer contains an icon that looks like a tall "H". This icon serves to indicate what region of the sequence is being displayed in the sequence widget (which usually presents less of the sequence, but in more detail). Scrolling the sequence widget causes this cursor to move in the map widget and, conversely, dragging the cursor around in the map widget will scroll the sequence widget. Once again, an underlying model facilitates this behavior. Both the sequence widget and the H cursor are views of an "interval" model that represents what part of the sequence should be displayed. Another instance of this kind of coordination appears in figure 4, in which a sliding control embedded in a chromosome widget controls the viewing area of a multi-map widget.

Additional Features

In addition to those features provided by the bioWidget architecture, the individual widgets themselves have a number of useful facilities. The sequence widget can display graphs computed dynamically from the sequence it is displaying; for instance, percent-GC content for DNA, as shown in figure 5, or hydrophobicity for polypeptides. It can also display the conceptual protein translation of a DNA sequence in any reading frame, using whatever genetic code is desired. A simple cut and paste facility allows sending a subsequence of interest to an external program for further analysis. Menus allow users to specify preferences for colors and fonts, and simple printing to a Postscript file is supported. Annotation displayed in the map widget can be classified into groups, for example, by organism, type of annotation, laboratory of origin, and so on. Groups can be shown or hidden to aid in filtering the available information.

Who uses bioWidgets?

Software can be reused at a number of different levels, by users with correspondingly varied needs. We intend to target three principle groups of bioWidget users. The first group of users consists of those who are looking for complete (but still component-based) solutions. To this group we provide standalone Java™ applets that can be embedded in web pages, taking input from files in standard formats (e.g. FASTA[15]) and providing sensible default behaviors. In this mode of operation the bioWidgets can be put to immediate use, but a user's configuration options are limited by what parameters can reasonably be encoded as HTML tags in a web page.

A user from the second group wishes to construct or customize an applet or standalone application, but without doing any Java™ programming. This is the JavaBeans™ user group, to which we provide a set of components defined according to the JavaBeans™ specification. The bean versions of the bioWidgets can be wired together to form an application using a visual development environment such as JavaStudio™[16]. See figure 6 for a simple example of such an application; in this case the user has drawn lines between the components to indicate that the database access bean should send sequence data from GSDB[17] to the sequence display bean, which in turn has been connected to a scrollbar bean. This kind of integration is possible because the bioWidget architecture extends the standards put forth in the JavaBeans™ specification.

Finally, the third group of users is comprised of Java™ programmers who wish to use the bioWidget architecture to achieve a goal that cannot be accomplished by connecting existing components. Examples include creating new widgets or reading data in a completely new format into existing widgets. These users are users not so much of the bioWidgets themselves, but of the underlying bioWidget architecture.

The bioWidget Consortium

The bioWidgets project has adopted a consortium model [18]. But, unlike groups such as the OMG[19] and the ODMG[20], the bioWidgets project is concerned with producing both specifications *and* implementations. Thus the consortium's administrative structure must facilitate and coordinate the release and versioning of both design standards and also the programs and applications designed according to those standards.

A Central Repository

The essential resource provided by (and to) the consortium is a central repository—both for finished components and software, and for technical proposals and/or ratified and evolving standards and specifications. The repository is responsible for versioning the bioWidgets, providing components to software developers and users, and tracking subsequent bug reports and requests for additional features. These reports and requests are in turn made available to anyone interested in working on existing widgets, whether or not they happen to be the original authors. By acting as a mediator in this fashion, the repository goes a long way towards guaranteeing the components' ease-of-use.

Quality Control

Nat Goodman's group at the Jackson Laboratory maintains the consortium's central repository, and has also assumed the responsibility of providing quality control for the entire endeavor. An essential part of any software sharing effort, the quality control process ensures that, prior to submission into the repository, each widget meets certain criteria with respect to documentation, engineering, and reusability.

Future Work

We will proceed to develop new widgets to support the display of a wider variety of data; future bioWidgets will handle data describing pedigrees, metabolic pathways, 2-D gels, gene expression arrays (SAGE, DNA chips, etc.), and database schemas. The development of particular widgets will soon exploit the upcoming support in web browsers for improvements in the Java™ language; these improvements address such issues as printing and cutting and pasting. For example, one will be able to select a sequence and paste it directly into a word processor document. Or one will be able to copy sequences from a number of web pages and paste them into a multiple alignment widget.

Finally, we hope to use the same facilities we have defined for inter-widget communication to enable communication between the widgets and a variety of remote data sources, in a more dynamic fashion. For example, one might highlight a region of sequence and request a BLAST[20] or motif search. The widget would communicate with a specialized remote server that would perform the requested task and return the results directly to the widget for display and possible subsequent analysis.

Using the widgets in this manner overcomes the main limitation of existing web-based systems, namely that, for the most part, they have no memory of what a user has done in the past, beyond what can be encoded in a URL. We also plan to expand the power of the widgets to act as interfaces for composing and answering queries on the data they display, rather than merely acting as browsers. This can be done by integrating the widgets with the kind of flexible query engine provided by systems like BioKleisli (see Chapter ??) and Multi-Database OPM (see Chapter ??).

Conclusions

The time has come for a component-based revolution in bioinformatics. The software technology, including the World Wide Web, Java™ and its diverse facilities, and other object-based component architectures such as CORBA will drive the effort. The growing abundance of data in need of analysis, the commonality of visualization needs across genomics applications and laboratory environments and the limits of developer resources will combine to create an intense "market" for GUI components. The bioWidgets Consortium will fill the necessary role of coordinator of widget development efforts and dispenser of widgets, and the bioWidget architecture will provide the technical backbone that ensures reusability and interoperability.

References

1. Goodman, N., Rozen, S., and Stein, L. (1995) "The Case for Componentry in Genome Informatics Systems". http://www-genome.wi.mit.edu/informatics/componentry.html
2. Searls, D.B. (1995) "bioTk: Componentry for Genome Informatics Graphical User Interfaces" Gene 163(2):GC1-16.
3. Java™ language. http://www.javasoft.com/

4. Gish, Warren (1994-1997). unpublished. Altschul, Stephen F., Warren Gish, Webb Miller, Eugene W. Myers, and David J. Lipman (1990). Basic local alignment search tool. J. Mol. Biol. 215:403-10.

5. Durbin, Richard and Thierry-Mieg, Jean. (1991). A C. Elegans Database (ACEDB). http://probe.nalusda.gov:8000/acedocs/acedbfaq.html

6. Stein, Lincoln and Thierry-Mieg, Jean. JADE. http://alpha.crbm.cnrs-mop.fr/jade/jade.html

7. Helt, Gregg. Genome Browser.
   http://fruitfly.berkeley.edu/javademos/GenomeBrowser.html

8. Grigoriev, Andrei. DerBrowser. http://www.mpimg-berlin-dahlem.mpg.de/~andy/DerBrowser/

9. MapView. http://www.gdb.org/gdb/mapviewHelp.html

10. Glenn E. Krasner and Stephen T. Pope. A cookbook for using the model-view controller user interface paradigm in Smalltalk-80. *Journal of Object-Oriented Programming*, 1(3): 26-49, August/September 1988

11. JDBC™ database access API.
    http://www.javasoft.com/products/jdbc/index.html

12. The Common Object Request Broker Architecture (CORBA).
    http://www.omg.org/

13. Java™Beans™ component architecture.
    http://www.javasoft.com/beans/index.html

14. GAIA. http://agave.humgen.upenn.edu/gaia/

15. W. R. Pearson and D. J. Lipman (1988), "Improved Tools for Biological Sequence Analysis", PNAS 85:2444-2448

16. JavaStudio™. http://www.sun.com/studio/

17. Genome Sequence DataBase. http://wehih.wehi.edu.au/gsdb/aboutgsdb.html The bioWidget Consortium homepage at Jackson Laboratory.
    http://goodman.jax.org/projects/biowidgets/consortium/index.html

18. Object Management Group (OMG). http://www.omg.org/

19. The Object Database Management Group (ODMG). http://www.odmg.org/

22 ACEDB: THE ACE DATABASE MANAGER

Jean Thierry-Mieg*, Danielle Thierry-Mieg*
and Lincoln Stein**

*Centre National de la Recherche Scientifique,
CRBM and Physique Mathématique,
1919 route de Mende
34293, Montpellier, France
** Cold Spring Harbor Laboratory
One Bungtown Road
Cold Spring Harbor, New York 11777 USA

Overview

The purpose of this paper is to review the main design problems involved in the development of a genome database system and to describe the ace kernel, the stand alone object oriented database manager underlying the C.elegans graphic "Acedb" program.

The ace kernel can handle large amounts of heterogeneous data with a complex evolving schema. It includes a query language and a basic graphic toolbox. It is optimized for speed, memory allocation and disk usage. It has efficient crash recovery and has been intensively tested over the years by demanding biologists. More recently the kernel has acquired client-server capabilities, concurrent write access, visibility over the network, a Java and a Perl interface.

The Ace kernel runs on any Unix workstation or PC with Linux; a port to Microsoft Windows is in the testing phase. It is freely and immediately usable by any interested party. Source code, binaries and documentation can be downloaded from *http://alpha.crbm.cnrs-mop.fr*

History

Acedb was designed by Richard Durbin and Jean Thierry-Mieg to manage and distribute genetic data on the nematode C.elegans. A survey at the end of 1989 had convinced the authors that an object oriented architecture was desirable but that no existing object oriented manager had the required capabilities to handle the complete C.elegans dataset. A new system was developed from scratch in the C programming language, a schema was chosen, the data was collated and the graphic C.elegans Acedb was first released to the worm community in June 91.

Because the source code was made freely available from the start[1], Acedb was rapidly adopted by communities working on a wide variety of organisms, including the plant Arabidopsis (Mike Cherry, Sam Cartinhour, then John Morris), the fruitfly Drosophila (John McCarthy, Suzanna Lewis, Frank Eeckman and Cyrus Hamon), the yeast S.cerevisiae (Mike Cherry), comparative vertebrates (Jo Dicks and John Edwards), edible plants (Doug Bigwood and Sam Cartinhour) and man (David Bentley and also the Integrated Genome Database (IGD) project, with Otto Ritter, Detlef Wolf, Petr Kocab, Martin Senger, Jaime Priluski). Acedb is also used in several non-biological applications, including astronomy and semiconductor manufacturing. A more complete list can be found in the Acedb FAQ (Frequently Asked Questions list), first assembled by Bradley Sherman in 1994, and now maintained by Dave Matthews [2].

A number of acedb workshops were organized, successively in Cambridge, Boston, Montpellier, San Francisco and Cornell. They last a couple of weeks and allow ample time for discussion and cooperative programming.

At the same time that it was spreading to other organisms, Acedb was becoming more important to the Genome Project. It has been used by the Sanger StLouis consortium, led by John Sulston and Bob Waterston, to manage data and annotate over 100 megabases of DNA sequences and more than 1 million expressed sequence tags (ESTs). As Acedb grew, we used its rich graphic interface for front ends to a number of specific analysis programs, allowing trained users to build maps, find genes (Phil Green) or display multiple alignments (Erik Sonnhammer). Direct live interfaces to external programs (Blast, the OSP oligos selection program of LaDeana Hillier, BioMotif of Gerard Mennessier and the Netscape Web browser) were also made available. A complete system for sequence assembly and editing, called Acembly, was built over Acedb and is distributed separately at http://alpha.crbm.cnrs-mop.fr/acembly.html

The development over the last few years of the client server architecture, described below, allows concurrent write access, overcoming the main limitation of acedb in a multiuser environment.

The fact that so many programmers and end-users were using Acedb had both good and bad effects. On the one hand, the speed and reliability of the database manager and of the basic graphic tool box continuously improved, as bugs were systematically chased and corrected in particular by Simon Kelley and Michel Potdevin. Detlef Wolf contributed a series of regression tests called "Aquila" to maintain code quality, and a commercial product called Purify was used to detect memory leaks. By 1996 the database kernel had reached a very stable state.

On the other hand, most users came to Acedb not for its underlying data management functions, but rather for the graphic displays contained in the X-windows version of Acedb called xace. The displays represent maps of the chromosomes at various scales (genes, clones, sequences), and allow users to jump from one map to the other. This is complemented by multimaps and multiple alignments, and other specialized data visualization tools, including a clone grid, a gel with restriction fragments and an image viewer. At the beginning, these displays looked very close to what most people wanted. But when each group tried to customize the displays to its particular needs or to add new displays, a number of difficulties became obvious. The graphic display code became always larger and more heterogeneous. Despite lots of efforts, some of the graphical displays are still buggy, some have complex and obscure user interfaces, and most remain undocumented.

The management problem stems from the fact that in the original Acedb design, the database manager, the biological graphic interface and the schema are intertwined. The individual displays often rely on particular features of the schema, creating rigidity and non-reusability. However, the schema, the graphics, and even some of the low level library calls were not chosen for maximum generality, but to fit the specific needs of the Nematode Genome project. Exacerbating this problem was the fact that Acedb evolved over time as the needs of the nematode project changed. As a result, the programs or variants that had been developed by other groups soon became incompatible with the newer versions of the code developed for C.elegans. This created frustration in the user community and several of the most active groups eventually dropped Acedb.

The lesson is the following: because we distribute the source code, people can recompile on any machine, fix bugs, and contribute new modules. This promotes rapid development, but alone is not sufficient. A centralized control of the source code is needed to keep the code base from diverging among a hundred mutually incompatible paths. Otherwise it is impractical to try to put the egg back together again by linking locally developed modules with the bulk of the distributed graphic Acedb.

To cope with this situation, we now propose two complementary solutions. On the one hand, we have redesigned most Acedb graphic displays for flexibility. They can now be reconfigured at the data level without touching the source code. On the other hand, we have developed a client/server architecture, with clients written in C, Java, or Perl.

We would like to stress that the key to the solution is to disconnect the data management schema from the application code. Otherwise, the application libraries, even if they are written as external clients, impose backward compatibility and progressively forbid any evolution of the underlying database. This problem is not unique to Acedb and our experience and partial solutions may interest the developers of other database systems.

The ace data definition language

Data in Acedb are organized in objects and objects are grouped in classes. Each object belongs to a unique class, and has a name which is unique in its class. Rather than having a fixed number of fields determined by their class, objects have a flexible internal structure, organized hierarchically as a tree. This tree contains basic data elements, such as numbers, character strings, and pointers to other objects, as well as identifiers called "tags" that give structure and meaning to the tree. The database schema consists of tree specifications, which in Acedb terminology are called class "models". A formal BNF definition and several examples and tutorials are available at [2]. Consider the following simple example:

```
?Person Paper ?Paper
        Address   Street Text
                  City   ?City
                  State  ?State

?City State ?State XREF Cities

?State Cities ?City XREF State
```

This schema defines models for three classes: Person, City and State, prefixed by question marks, and six tags: Paper, Address, Street, City, State and Cities. Inside trees, the symbols "?classname" represent pointers to objects in the specified class. The XREF tells the system to maintain automatically the cross referencing between State and Cities.

Note that in class Person, under the tag Address, we have three tags on equal footing: Street, City and State. This nesting brings several benefits. First, it remains clear when the model becomes complicated. Second, this construction provides an automatic clustering of the contents. Any person for whom State is specified automatically gains a partial Address and can be selected and retrieved on this criterion allowing natural queries and automatic classification. Finally, this provides a very simple way, explained below, of de-correlating the methods from the schema.

The tag[2] system

The greatest problem we encountered with the early versions of Acedb was a tight coupling of the display to the schema, which has come to be known in the Acedb community, particularly among data curators, as the "magic tag" syndrome. The meaning is this: features presented in the applications and in the graphic interface were obtained from the database by reference to explicit tags in the class definitions.

Consider a simple example where one wants to write a program for addressing envelopes. The easy way is to look at the definition of class Person, to retrieve the Street, City and State values and print them. But this piece of code turns the tags

Street, City and State into "magic tags". If a new curator defines his own schema and replaces the tag State by the tag Country, this addressing method will no longer work. As more and more methods are defined using class tags, the schema becomes totally frozen and the database loses its versatility.

The solution came through discussions with Otto Ritter. Rather than accessing the particular tags Street, City, State, we access the tag Address and collect everything two to the right. We call this the tag[2] system (pronounced "tag-two"). Although deceptively simple, this method does provide the desired decorrelation between the data and the methods. Indeed, this new mailing method will now correctly send mail to a sailor with a different sort of address:

```
?Sailor Address  Ship ?Ship
                 Harbor ?City
                 Country ?Country
```

Constructed types

The nesting of tags under tags can be generalized and Acedb allows recursively constructed types, indicated by specifying a model at a leaf position of another model. This can be used either to specify named attributes of a relation or to define an arbitrarily long list. Consider the following involved example:

```
?Ship    Cargo ?Goods  #Cargo_info
         Route #Route_element

#Cargo_info Owner ?Owner
            Weight UNIQUE Float
            Value UNIQUE Float

#Route_element ?City Int #Route_element
    // The Int number represents the number of days
```

The cargo information is not directly a qualifier of the Ship, but rather an attribute of the Ship/Goods relation, allowing to say that Tom owns 60 tons of bananas aboard a given ship. The route information, which is recursively defined, will consist of an ordered list of cities and travel delays.

In general, tags are multivalued, i.e. the Cargo of a given ship may consist of several Goods, and a given piece of Goods may have several owners, but the modifier UNIQUE ensures that each object has at most one Weight and Value.

Comments may be attached to any data element. They do not interfere with the query language or the application routines, but they are indexed, which allows direct retrieval of the objects containing them. In addition, every data element which is added or modified is automatically time stamped with the date and the name of the user. By the way, dates in Acedb are not subject to year 2000 problems!

The ace file format

When entering data, Acedb can be thought of as a data compiler. One prepares text files, following a particular format called .ace (pronounced dot ace) format. One submits them to the Acedb parser and the data are compiled into an internal binary representation which allows the kernel to retrieve them easily when needed. Any error in the data files, i.e. any data line which does not fit the schema is reported with an error message and line number. One can fix the data file and read it again until there is no error, in very much the same way as one would write and compile a program in Fortran or other programming language.

In fact Acedb can be used simply as a high level syntax checking program. One writes the schema, and parses the Acedb files to detect wrong tag names and ill-formed data and then discards the compiled database and forwards the ace files to some other system. An example of this is Common Assembly Format, which are .ace files used in the Sanger and St Louis sequencing projects to exchange data across assembly programs [3]

The following few lines constitute a valid ace file for the shema given in the preceding paragraph:

```
Person Tom
City London

Ship Enterprise
Cargo Lemons Value 6732.5
Route Tangier 5 Bordeaux 8 "New York"
// Quotes are needed around New York
// to protect the blank space

Person Tom
State England
```

Acedb is cumulative, and different pieces of information on a given topic can be entered in different paragraphs, as above for Tom, or even in different ace files. At any time, the data can be re-exported in the same format. Acedb also allows data to be exported as tables for use in a relational database or other program.

An important effect of the syntax is the ability to exchange data between databases with different schemata. The first line of each paragraph is a class name. Then each line starts with a tag which is recognized by considering the model of that class. The intermediate tags, address in this example, are not needed in the ace file, they are reconstructed by the system. Hence, if in the target database, the tag Street is located in a different position, the data can still be read.

Furthermore, it is possible to edit the schema of the database without losing the current data, so that when one receives some new type of information one can add it to a running database after editing the schema in an adequate way. Acedb curators use this method and usually distribute schema modifications with each new data release.

Comparison with other databases

When comparing Acedb with other systems one should bear in mind that today two broad categories of database are in use. Relational systems, like Sybase and Oracle, organize the data in tables, a little like phonebooks. Object oriented systems like O2, ObjectStore, OPM, Matisse or Acedb organize data in clumps, like a patient record kept by a physician. Relational systems are best when the schema is simple, the data are regular and successive queries are independent. Object systems are best when the schema is complex, the data irregular and the queries correlated.

In biology and any other experimental science, one knows a lot in a few cases, and little in most others. This makes relational systems inefficient since one has to maintain and explore many quasi empty tables. In addition, one often wishes to start somewhere and progressively explore the surrounding area. However, relational systems do not have the concept of neighbors, so these local explorations are not automatically optimized. Whereas in object systems small objects cost little and the direct links between related objects, create a natural   topology.

These two characteristics thus favor, in experimental sciences, the choice of an object oriented system. On the other hand, relational systems, which are very well adapted to commercial activities, are well supported, offer the advantages of industry standardization and the ability to handle large numbers of concurrent clients. For these reasons many attempts at adapting relational systems to represent biological data have been made.

We think that it is fair to say that, when the schema is complex, relational databases are much more difficult to maintain than object-oriented systems. A case in point is GDB, which, after many years of maintaining a complex relational schema, ultimately migrated to a hybrid system: the schema is defined in the OPM object-oriented data definition language but the data are stored in Sybase. Performance does suffer from this union, but the system designers felt that the gain in maintainability was worth the loss in performance.

The data definition languages of the object oriented systems, O2, OPM, ObjectStore and Acedb are very similar. There is however a subtle difference in the way the data are considered. In Acedb, the objects are fully exportable using our generic data exchange system, the .ace format. In these other object-oriented databases, a similar tool is missing, not because it would be so difficult to develop, but because in these systems, objects have a hidden internal identifier which cannot be exported. Two objects in O2 can have all their fields identical and yet be considered different, whereas in Acedb an object is completely identified by its name and class.

Finally, Acedb is easy to configure because the schema of the database is itself treated as an object. This implies that it is possible to refine or modify the schema of a running ace database without losing the existing data. It is not necessary in Acedb to design *a priori* the complete schema. Rather, one can start with a minimal canvas and extend it when more data become available. The design of the database is progressive and data driven.

Yet another interesting system is the Boulder input/output language defined by Lincoln Stein [4]. A boulder file is very similar to an ace file, but without any underlying schema. It is divided into paragraphs called "stones", separated by a record delimiter character. Each data line starts with a tag name followed by some data, in a format that can be parsed without a knowledge of the data semantics. The system is designed to allow programs to be piped together. Each program accepts and emits a stream of Boulder stones. As stones flow along the pipe, each program is free to extract or modify the tag lines it is interested in and reexport the stream. The data repository is the Boulder stream itself. It has no schema, no exact boundary, and no clearly identified hardware localization. In fact, Boulder is not a database system, and does not support queries, but it is very light and most convenient in a production environment. Boulder and acedb are highly compatible, since an ace file is a valid Boulder file and an ace model can always be constructed by inspection of a set of boulders.

The query language

In the first Acedb release in 1991, one could only search for a given word and then browse through the graphic hypertext and map displays. Acedb browsing is faster but similar to the more recent Web user interface originally defined by the Mosaic Web browser. It is easily picked up by end users. However, one often needs to ask specific questions.

Relational databases offer a standard query language called SQL, but in 1991, there was no equivalent for object oriented databases. Therefore, we developed and implemented in Acedb a simple language allowing the selection of subsets of objects satisfying certain properties and navigation to their neighbors. For example one can select the prolific authors with the query:

```
Find author COUNT paper > 100
```

Or the alleles of the genes on chromosome III with:

```
Find Gene Map = III ; Follow Allele
```

To write those queries, one needs to learn the syntax and the schema of the database. For example, and this is not intuitive, one must know that in the C.elegans database the class Map refers to the genetic map of the chromosome. To facilitate the composition of queries, John McCarthy, Gary Aochi and Aarun Aggarval added to the Acedb graphic interface two helpful tools, Query-builder and Query-by-example which prompt the user for the correct fields and the correct syntax. Once written, the queries can be named, saved, reused and may contain positional runtime parameters.

This language, which acts like a SQL Select statement, is complemented by Table-Maker, a graphical tool that leads the user through the steps of creating a tabular view of a portion of the database. Table-Maker allows the gathering of data from multiple object classes into a summary report for the purposes of visualization or data exportation. The table definition can then be saved and reused. In graphic mode, the complete table must be held in memory, and on normal machines this limits the length of the returned table to a few thousand lines. But after several rounds of optimization, the current acedb_4.5 release can export in text-only mode arbitrarily long tables without using much memory. Release 4.7 will incorporate a fuzzy logic layer which greatly accelerates the selection of rare objects and is the result of many years of discussions with Otto Ritter.

The present system is efficient, expressive and well tested, but it is non-standard, and the table definitions are hard to compose outside of the graphic interface. In the meantime, the O2 group has developed the object query language OQL[5] which was adopted as a standard by the Object Definition Management Group ODMG. OQL is a large system, with nice features like the dynamic construction of new objects, but it is often clumsy; for example the syntax of a query differs if a field is single or multivalued, and OQL does not support well the recursivity of the Acedb schema. Other proposals, like Lorel [6], are closer to Acedb. Following those ideas, Richard Durbin started last year the design of a new query language for Acedb, which may be available by the end of 1998. This should simplify the querying of the database from external clients and allow more general questions.

Client server

Acedb is mostly used as a standalone single user program, and indeed it runs well even on a low-end system. For example, the full C. elegans database is very useable even on a 133 MHz PC laptop running Linux. However, we have also developed a TCP/IP-based client/server architecture intended for heavy multi-user applications and Web-based interfaces. The server runs as a Unix network daemon, listening for requests from incoming clients, and replying with textual or graphical representations of objects (the latter are returned as binary GIF images). Data can be imported into the server locally or remotely in the form of .ace files, or bulk-loaded from FASTA files (nucleotide and protein sequences only).

Several clients are defined: a simple dialog interface called aceclient, which has a very small overhead and is convenient to use interactively over a slow network or to embed within a shell script; a simplified restricted interface called jade2ace, which we use to communicate with Java clients, a complete Perl interface called AcePerl which is convenient for scripting, and finally a fully graphic client called xclient, which is functionally equivalent to the graphic Acedb code but gets its data transparently from a remote server.

Xclient must be installed on the client machine. When the xclient is running, data is imported on the fly from the distant server and used to construct a local database. The resulting system is fast, even on a slow network, because most actions operate on locally cached data. Xclient can be used in two ways. Inside the lab, it allows multiple clients to get simultaneous write access on the server, a feature that was lacking in xace. Over the network, it allows users to access the latest data without the need to download a complete data set. Xclient may be downloaded from our Web site. By default, it is configured to access our C.elegans server. Instructions to redirect the client or to create a new server are included.

Web interface

The standard graphic acedb, xace, uses the X11 protocol. It does work over the network but only by starting a big process on the server side, which is costly and requires special privileges. Xclient, described above, needs to be installed before it can be used and only runs on Unix machines. In this section, we explain how Acedb databases can be viewed over the Web using conventional Web browsers, without any special effort from the end user.

The first implementation of Acedb over the Web was written as early as 1992 by Guy Decoux. Called the Moulon server, it was surprisingly complete but very slow, because a new Acedb process was started to process each request.

When the aceserver became available, Doug Bigwood's group at the National Agriculture Library–developed Webace. We wrote together a special module to export Acedb objects in a form that could be compiled into Perl objects by the Perl interpreter. These objects were then enriched outside of Acedb by other-Perl layers that add hot links. For example, GenBank accession numbers are automatically converted into links to NCBI Entrez pages. Webace also allows several Acedb databases to be browsed simultaneously. The downside of Webace is that installation is somewhat complex; however it is very powerful and the NAL Webace server is actively used to access data concerning a large number of cereals and other interesting plants. Webace is now maintained by Tim Hubbard http://webace.sanger.ac.uk.

Two years ago, we started with great enthusiasm to work on a Java interface to Acedb. We designed a system called Jade [7] which may be used to browse different types of servers, including Acedb databases, relational databases, or even simple flat files. The low-level interfaces that allow Acedb objects to be converted into native Java objects are complete and quite useable, but the development of a graphical user interface to replace the xace displays has lagged behind. This is partly because of the immaturity of the Java abstract windowing toolkit (AWT), and partly because we were waiting for the Biowidget Initiative to provide a ready-made toolkit of graphical displays. Now that Biowidgets has faltered, we have resumed development of Java-based interfaces.

This year, we wrote a new Perl client library called AcePerl. It is well documented and easy to use and provides a simple way for a Perl script to read, modify and write the data contained in one or several aceservers. In conjunction with the CGI Perl module, an earlier work from Lincoln Stein, AcePerl provides a simple way to browse Acedb data from a personal Web page.

The remaining question is the construction of a real graphic page on the Web. The first approach is to directly export a graphic from acedb. The original Moulon server asked for a postscript print out and turned it into a gif file. This was simplified in Webace when Richard Durbin used the gif library from Tom Boutell to directly export gif documents from acedb, a feature now incorporated in the aceserver. These images can then be displayed by Webace, AcePerl or Jade.

However, the Acedb graphic displays with their popup menus, linked windows and complex color codes are not well adapted to the Web, and interactions with the server remain painful across a slow network. The solution is to do more processing on the client side and we believe that the long term solutions is to write all the displays in Java, once that language has fully stabilized, or in Perl, if a client-side version of that language embedded into Web browsers becomes widely available.

Portability

Acedb is fully portable to all Unix machines, including PC/Linux and 64 bits processors, like the alpha and mips chips. Some versions of the code were successfully ported on the Macintosh (Frank Eeckman and Richard Durbin in 1992, then Cyrus Hamon from 1994 to 1996). Finally, a Windows NT and Windows 95 version is under development by Richard Bruskiewich and is available as beta test software on the Acedb sites. A database constructed on one platform can be used by an executable running on another one.

Acedb is written in standard ANSI-C. Unfortunately, although things have been improving over the years, this is not sufficient to insure portability. Most compilers are not completely POSIX compliant and the system calls to read the disk, traverse the file system, to run external subprocesses or construct a client/server network connection diverge slightly but significantly. To circumvent this problem, we carefully isolated from the start of the development of Acedb all the potentially system-specific calls in a small collection of functions which we used throughout the code, and we replaced everywhere the inclusion of the standard C library "stdlib.h" by the inclusion of our own library of wrapper functions, allowing system calls to be normalized across unusual machines. The result continues to be very rewarding.

1) Acedb can be adapted and recompiled on any Unix platform in a few hours.

2) When the first 64 bit machine appeared, the Dec alpha, we only had to adjust a few low lying routine concerned with word alignment.

3) The port to the Mac and to Windows was greatly facilitated.

4) In the rare cases where some system calls give incorrect results on a given machine, we can easily work around the problem by touching a single file.

An independent problem is the binary compatibility of the database. As you recall, Acedb is a data compiler. It reads ace files and turns them into a binary database file structure. Different machines have different word alignments and different byte order (little and big endian). Simon Kelley solved this problem by introducing an optional byte swapper inside the disk access routines which works the following way. When you create a database with a given machine, acedb accepts the convention of that machine. Thus if you always use the same machine, acedb never byte swaps. However, if now you access the same database from a different machine with the opposite convention, Acedb automatically detects the need to byte swap and will do it transparently. The net result is that an Acedb database can be NFS accessible from an heterogeneous set of computers, which would not be possible otherwise. In fact, a Windows executable can read and edit a Unix created database.

Performance

In day to day use, Acedb usually seems very fast relative to other database systems, even commercial ones. This statement is hard to quantify because it is notoriously difficult to make comparisons among databases that vary widely in their schemas, design and implementation. Furthermore, it is quite difficult to compare relational to object-oriented databases, because there is no common measure concerning queries, data insertion rate or transaction throughput that can be applied uniformly to both types of systems. For this reason, we will try to give a subjective assessment of Acedb speed through a series of anecdotes.

Acedb is naturally optimized to handle point and browse queries that account for more than 95% of its use. However, when dealing with very large objects, acedb can occasionally be very slow, but no more than a relational system when forced to make a join across the dozens of tables required to represent an object hierarchy. Note that huge objects in acedb are in general the result of an inadequate design of the schema. For instance, in the example above, we have cross referenced State and City but not Person and City. Otherwise, a given city, say London, could point to a hundred thousand Persons and become very heavy to manipulate. In such a case, the adequate way to maintain a dynamic list of the London dwellers is to use an acedb subclass.

The C.elegans data set contains half a million Acedb objects and can be loaded on a pentium-200 PC running Microsoft Windows in around one hour. On a similar machine running Solaris, we can export from the St Louis Merck EST database a table of 13 columns and 1 million lines in 16 minutes. For the IGD project, we were able to load on a DEC alpha a large data set consisting of GDB, GenBank, SwissProt and several dozen smaller databases into acedb. Once loaded, Acedb was able to complete a test suite of queries on this data set in 3 minutes as opposed to 30 minutes for a Sybase server on the same machine.

It is crucial, however, to run the database on a local disk. If the database disk is mounted over NFS, or on a RAID system, performance easily degrades by a factor of 10. In such a case, to produce a very large table, it is much faster to ftp the whole database onto a local disk, run the table query and destroy the database, than to run the query off the remote disk.

How does the storage capacity of Acedb compare to other database management systems? Comparisons with relational systems are difficult because of the impossibility of relating relational tables to objects. However, we can compare Acedb to other object oriented systems. The present C.elegans data set contains 500 thousand objects and uses half a Gigabyte of disk space. Direct comparisons of storage capacity by counting the number of objects may be misleading because it depends on the expressive power of the grammar. For example, Gilles Lucato and Isabelle Mougenot at LIRMM wrote an automatic Acedb to Matisse translator. Extra classes were needed in Matisse to store the Acedb structured tags and as a result, the C.elegans dataset in Matisse is spread over several million objects and takes several times more disk space. Illustra also uses more diskspace. We have been told that O2 performance degrades at around 50 thousand objects on similar machines. ObjectStore databases are memory mapped to disk, this places a strict limit of 4 gigabytes on conventional 32-bit architecture machines. Acedb has no such hard limit, but typically one needs 50 bytes of memory and 1 kilobyte of disk space per object, effectively limiting acedb to one million objects and 1 gigabyte of disk on a 128 Mb machine.

How to get the software

The whole Acedb system, including the Java and Perl tools and demos, and the Acembly package are available from our Web page *http://alpha.crbm.cnrs-mop.fr.*

The C.elegans data and Acedb source code may be downloaded from
ftp://ncbi.nlm.nih.gov/repository/acedb/ in the US
ftp://lirmm.lirmm.fr/pub/acedb/ in France
ftp://ftp.sanger.ac.uk/pub/acedb/ in England

Documentation, tutorials and examples are maintained by Sam Cartinhour and Dave Matthews on the US site *http:probe.nalusda.gov:8000/acedocs/index.html*

Conclusion

The Ace kernel is a versatile database manager. It is fast, easy to configure and has a powerful grammar. Since the source code is available, it can be recompiled on any platform and many people have contributed modules and corrections. The client server architecture and the Perl and Java interfaces make it suitable for any application where the schema is complex and the data heterogeneous. The Ace kernel can be used independently of the graphic interface.

We believe that the Ace kernel is sufficiently fast and robust to support the human genome sequencing project as well as it has supported the C.elegans project and plan to continue to improve its speed and quality over the next years. We also welcome innovative uses for Acedb outside the biological community.

Acknowledgments

This work was supported in part by the European contract IGD-GIS/BMH4-ct96-0263.

We are greatly indebted to Richard Durbin for years of collaboration, and to all the people who helped us to write to test and to document the ace system, those we quoted, and those we did not. But we want also to thank the many users who for years have endured all the bugs, complained little, suggested a lot and suffered much, in particular LaDeana Hillier and Simon Dear.

[1] ftp://ncbi.nlm.nih.gov/repository/acedb

[2] http://probe.nalusda.gov:8000/acedocs

[3] Simon Dear et al. Sequence assembly with CAF tools, Genome Research 8, 260-268, 1998.

[4] http://stein.cshl.org Lincoln Stein et al., Splicing Unix into a genome mapping laboratory. Usenix summer technical conference, 221-229, 1994

[5] R.G.G Cattell, The object database standard: ODMG-93, MorganKaufmann, San Francisco, California, 1994

[6] Serge Abiteboul et al., The Lorel query language for semistructured data. Journal on digital libraries, 1, 23-46, 1997.

[7] Lincoln Stein et al., Jade: An approach for interconnecting bioinformatics databases, Gene, 209, 39-43, 1998.

23 LABBASE: DATA AND WORKFLOW MANAGEMENT FOR LARGE SCALE BIOLOGICAL RESEARCH

Nathan Goodman *, Steve Rozen **,
and Lincoln Stein ***

* The Jackson Laboratory, Bar Harbor, ME
** Whitehead Institute Center for Genome Research,
Cambridge, MA
*** Cold Spring Harbor Laboratoy, Cold Spring Harbor,
NY

Introduction

Laboratory informatics is an essential part of most large scale biological research projects. A key element of laboratory informatics is a *laboratory information management system* (LIMS) whose purpose is to keep track of experiments and analyses that have been completed or are in-progress, to collect and manage the data generated by these experiments and analyses, and to provide suitable means for laboratory personnel to monitor and control the process.

While information management systems are ubiquitous in modern commerce, the demands of biological research pose different and difficult challenges [1-3]. Biological research projects entail complex, multi-step procedures (called "protocols") typically with multiple branch points and cycles. Such protocols involve a combination of laboratory procedures and computational analyses. The former may be carried out manually by laboratory personnel, or in an automated fashion using robotics or other forms of instrumentation, or (most often) by a combination thereof. Computational analyses may be carried by programs developed internally, or by ones supplied by academic or commercial third parties, or (increasingly) by public or proprietary Web servers; some analyses are simple and fast, while others may consume hours or days of computation and may require

280

dedicated computing equipment. Special cases requiring atypical, one-of-a-kind treatment are common. Laboratory procedures inevitably fail from time-to-time, and graceful handling of failures is essential. The data types themselves are complex and may be related in complicated ways. Superimposed on this intrinsic complexity is a rapid rate of change: research laboratories are staffed by highly trained and motivated scientists who continuously strive to refine old techniques and invent new ones in the never ending struggle to do the best possible science. Frequently, old and new methods must coexist as new methods are developed and prove their worth. A not infrequent occurrence is for a promising new method to enter service only to be retracted as unforeseen problems arise.

We have been pursuing a strategy for laboratory informatics based on three main ideas [4-7]: (1) component-based systems; (2) workflow management; and (3) domain-specific data management. The basic ingredients of our strategy are application programs, called *components*, which do the computational work of the laboratory project. Components may be complex analysis programs, simple data entry screens, interfaces to laboratory instruments, or anything else. We combine components into complete informatics systems using a workflow paradigm [8-10]. It is natural to depict such systems as diagrams (cf. Figure 1), reminiscent of finite state machines, in which the nodes represent components, and the arrows indicate the order in which components execute. We refer to such systems as *workflows*. As a workflow executes, we store in a database the results produced by each component, as well as status information such as which components have executed successfully, which have failed, and which are ready to be scheduled for future execution. We use specialized data management software to simplify this task.

This article describes the workflow and data management software we have developed pursuant to this strategy, called LabFlow and LabBase respectively. LabFlow provides an object-oriented framework [11] for describing workflows, an engine for executing these, and a variety of tools for monitoring and controlling the executions. LabBase is implemented as middleware running on top of commercial relational database management systems (presently Sybase and ORACLE). It provides a data definition language for succinctly defining laboratory databases, and operations for conveniently storing and retrieving data in such databases. Both LabFlow and LabBase are implemented in Perl5 and are designed to be used conveniently by Perl programs.

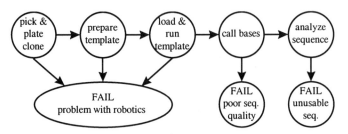

Figure 1: Sample Workflow

We will use as a running example a hypothetical laboratory project whose purpose is to sequence and analyze a large number of cDNA clones drawn from several libraries. A database for such a project would store information about (i) the libraries being sequenced, (ii) the clones picked for sequencing, (iii) sequence-reads performed on those clones, (iv) assemblies of sequence-reads (to coalesce multiple reads from the same clone, and to detect and exploit situations in which duplicate clones are picked), and (v) analyses of assembled sequences. The system would include many components, including (i) software to control robots involved in clone-picking and preparation of sequencing templates, (ii) base calling software, (iii) software to strip vector and for quality screening of raw sequences, e.g., to detect *E. coli* contamination and check for repetitive elements, (iv) sequence assembly software, (v) sequence analysis software, and (vi) some means for laboratory personnel to review the results.

LabBase and LabFlow are research software and are incomplete in many ways. We will endeavor to point out the major holes that we are aware of. The software is freely available and redistributable (see http://goodman.jax.org for details).

LabBase Data Management System

LabBase provides four main concepts for modeling laboratory (or other) databases: Objects, Materials, Steps, and States. *Objects* are structural objects, similar to those found in ACEDB [12], OPM [13], lore [14], UnQL [15], and many other systems. *Materials* are Objects that represent the identifiable things that participate in a laboratory protocol, such as libraries and clones. *Steps* are Objects reporting the results of a laboratory or analytical procedure, such as sequencing a clone, or running BLAST [16] on a sequence. *States* are Objects that represent places in a laboratory protocol, e.g., "ready for sequencing" or "ready for BLAST analysis". We use the term *object* (lower-case) to refer to any kind of Object including a Material, Step, or State.

The most compelling feature of LabBase is that it provides built-in support for two relationships among Materials, Steps, and States that lie at the core of typical laboratory databases. One is a relationship connecting Steps to the Materials upon which they operate. When a Step is stored in the database, LabBase automatically links the Step to its operand Materials in a chronological history and provides a means to access Step-data directly from these Materials; for example, one can retrieve a clone's sequence or a sequence's BLAST analysis by querying the respective Materials rather than the Steps. The second built-in relationship connects Materials to States. When a Material is created, LabBase provides a means to place the Material in an initial State; then as Steps operating on the Material are created, the system provides a means to move the Material to the appropriate next State thereby tracking its progress through the protocol. Both of these relationships are many-to-many. We discuss these relationships further when we describe LabBase operations.

To create a database for a specific laboratory protocol, the main tasks are to give names to the Materials and Steps of interest, and to describe the data to be reported in each Step.

Steps are generally obvious, because they correspond to the actual work being done in the laboratory protocol. The main subtlety is ensuring that Steps correspond to useful points of contact between the laboratory and the computer. In our running example, possible Steps include ones reporting (i) that a library has been constructed, (ii) that a clone has been picked and plated, (iii) that sequence-template has been prepared from a clone, (iv) that sequence-template has been loaded onto a sequencing machine and run, (v) the results of a sequencing-run, e.g., base-calls, quality indicators, and chromatographs, (vi) the results of vector stripping and quality screening of sequencing results, (vii) the results of assembling sequences, and (viii) the results of analyzing sequence-assemblies.

Many Materials are equally as obvious, because they correspond to the major reagents employed in the protocol, e.g., libraries and clones, or the major data produced by the protocol, e.g., sequence-reads and assemblies. As with Steps, the main danger is excess: Materials should only be defined for things that are really worth tracking. Limitations in our current software push strongly in the direction of parsimony. The mechanism mentioned above for connecting Step-data to Materials only works for Steps operating directly on a Material; it does not work transitively over related Materials. While it is easy to get the base-calls for a sequence-read, and a list of all sequence-reads for a given clone, and a list of all clones picked from a library, the software offers no special help for getting all base-calls for all sequence-reads for a given clone or library. A second limitation is that LabFlow (see later section) only supports workflows in which a single kind of Material marches through a protocol. The effect of these limitations is to encourage database designs in which multiple real-world material are elided into a single database-Material. In our example, it would probably be best to represent libraries as Objects (not Materials), and to merge clones and sequence-reads into one Material; assemblies would probably remain as separate Materials. The end result is a database with just two kinds of Materials: sequence-reads and sequence-assemblies.

To recapitulate, the database for our running example would have just two kinds of Materials, sequence-reads and sequence-assemblies, and many kinds of Steps, each operating on one Material. One of the possible Steps listed earlier, namely, the one reporting on library construction, must fall by the wayside, since we have decided to represent libraries as Objects, not Materials; data on library construction would be stored as fields of these library Objects. The most obvious, practical shortcoming of this example database is that without a clone Material, we lose the most natural means of coordinating multiple reads from the same clone. In the database as given, one would probably coordinate multiple reads per clone in the context of sequence-assemblies; this may be workable but is certainly not ideal.

LabBase Details

A LabBase object is a collection of fields. The data that may be stored in a field include numbers, strings, nucleotide and amino acid sequences of arbitrary length, sub-objects (called *structs*), lists, pointers to other objects (called *object-references*), and several others. Objects, Materials, and Steps may not be stored in fields, although references to these elements may be stored. LabBase makes note of the time each object was created and who did the creation, and stores this information in special, predefined fields of the object. Every LabBase object has a unique *internal object-identifier* assigned by the system when the object is created; the system uses this identifier to locate objects in the database and to connect objects together. An object may also have an *external object-identifier* assigned by the client program; an external object-identifier is similar to a primary key in a relational database.

LabBase objects are strongly typed: the system knows the type of every object, the fields that may appear in the object, and the data types of the values that may be stored in each field. A field may be *mandatory* or *optional* for a given type of object. An optional field which is absent from an object is said to be NULL. Fields are single-valued, meaning that each field contains a single value of its allowed type (or perhaps no value if the field is optional). This is in contrast to the multi-valued fields of ACEDB. Lists may be used to emulate multi-valued fields; it is possible to specify that empty lists be treated as if they were NULL which is useful when using lists for this purpose. LabBase provides no explicit support for compound attributes (e.g., pairs of x, y coordinates), but structs can be used to emulate this feature; it is possible to specify that empty structs be treated as NULL which is useful when using structs in this manner. Field-names are global, which means that if fields in two or more object-types have the same name, they must also have the same type. This is unlike most relational databases, where column-names are local to each table. LabBase adopts this unusual property to facilitate retrieval of Step-data via Materials as mentioned above and discussed further below.

A LabBase object is essentially a hierarchical record which is "object-like" in that it has a unique identifier. LabBase is far from a full object-oriented database [17] system lacking, among things, (i) support for inheritance or class hierarchies in any general sense, (ii) user defined data types, and (iii) database support for methods (i.e., functions) applied to objects.

Technically, Object, Material, and Step are *type constructors*, not types. These constructs are used to define types of objects, such as "clone". Having done so, one can then create actual objects (instances) of the types, e.g., an object representing a specific clone. We say colloquially that a clone is a Material, but it is more precise to say that a clone-object is an instance of the clone-object-type, and that the clone-object-type is a kind of Material. We will avoid this circumlocution whenever possible. By contrast, State is an actual Object; LabBase provides a predefined object-type, called -state, and a State is simply an instance of that type. (By convention, all system-defined names, such as -state, start with '-').

LabBase supports a "fetch-and-store" database interface, also known as a "two-level store", similar to most relational databases [18]. Objects must be explicitly stored in the database to create them and explicitly fetched from the database to retrieve an up-to-date copy. Updates to an object must also be explicitly stored back in the database.

The workhorse operations in the system are put which creates and stores objects in the database, and get which retrieves objects or selected fields from objects satisfying a condition. The "conditions" that can be specified in get (and other LabBase operations described shortly) are far simpler than in many databases: an operation can manipulate all objects of a given type, or all Materials of a given type in a given State, or a single object denoted by its internal or external object-identifier; this is an area for future improvement.

The put and get operations can be applied to Objects, Materials, and Steps, but different options are legal depending on the type of operand; the various options have been arranged to provide exactly the capabilities needed by LabFlow. When creating Materials, put can place each Material into one or more initial States. When creating Steps, put can be provided with a list of Materials upon which the Step is operating. The system links the Step to each Material in a chronological history, and can be instructed to move each Material from one State to another. When retrieving Materials, get can move each Material to a new State. This is a convenient way to indicate that the Materials are actively being processed by the client program and to avoid anomalies that can result from two program processing the same Material at the same time; this technique is sometimes called "application-level locking". When retrieving Materials, all fields of all Steps on the Material's history are treated as virtual fields of the Material itself and may be retrieved by get exactly as if they were real fields of the Material. One important detail is that if the same field-name appears in multiple Steps on a Material's history-list, get retrieves the most-recent one; in particular, if the same kind of Step operates on a Material twice, e.g., if a clone is sequenced twice, get retrieves the results from the most recent application of the Step. In addition, get can retrieve two other virtual fields maintained by the system, namely, the list of all Steps that have operated on the Material (called its *history-list*), and the list of all States in which the Material currently resides (called its *state-list*).

LabFlow uses these operations in a specific pattern illustrated in Figure 2. First, it uses put to create a Material and place the Material in its initial State. Then, as laboratory or analytical procedures are performed on the Material, it uses get to retrieve the fields of the Material needed to perform the procedure; these are typically virtual fields obtained from Steps on the Material's history-list; the get operation also moves the Material to an "in-process" State (not shown in the figure) so that other incarnations of the same procedure will not try to process the same Material.

When the procedure completes, LabFlow uses put to store its results in the database as a LabBase Step, and to move the Material to its next State.

In addition to get and put, LabBase provides the following operations: count returns the number of objects satisfying a condition; delete removes from the database a set of objects satisfying a condition; update changes the values of selected fields in an object or struct; set_states changes the States of Materials satisfying a condition.

Every LabBase operation executes as an atomic transaction. For applications requiring more control over transaction boundaries, the system provides operations to begin, commit, and rollback transactions.

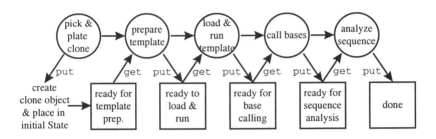

Figure 2: Interactions Among LabBase Operations, Materials, Steps, and States.

LabBase Summary

The facilities provided by LabBase make it easier to create a laboratory database by automating commonly occurring database design tasks. Since Steps are automatically connected to Materials, the database designer need not be concerned with how these links are maintained. Since Steps are automatically preserved in a chronological history, the designer need not be concerned with preventing updates from overwriting previous data. Since Step-data can be queried via Materials, the designer need not be concerned with storing laboratory results inside Materials, nor with providing views for this purpose. Since objects can contain lists and sub-objects, the designer need not be concerned with creating "link-tables" or similar devices to implement these structures.

It would be useful to extend the system to handle commonly occurring relationships involving Materials. Our example illustrates several situations in which one Material is *derived from* another; it would be useful for the system to "understand" such relationships and to propagate Step-data from a derived Material to its source and vice versa. *Grouping* is another common case, such as when a set of samples are arrayed in a plate or on a chip. Part/whole relationships are common also.

LabFlow Workflow Management System

LabFlow provides a Perl5 object-oriented framework for describing laboratory (or other) workflows, The most significant classes in this framework are Worker, Step, and Router. *Workers* encapsulate the components which do the computational work of the information system. The Worker class defines interfaces that allow arbitrary programs to be connected to the workflow. *Steps* in LabFlow are analogous to those in LabBase, but whereas in LabBase a Step merely records data, in LabFlow a Step embodies the computation that generates the data, and the "glue" that connects the computation to the rest of the information system. (The LabBase and LabFlow Step concepts are so similar that we find it natural to use the same term for both). Each Step has two parts: an input queue (actually a State) containing Materials that are waiting for the Step to be performed, and a Worker which does the actually computation. Technically, a Step need not contain a Worker or may employ multiple Workers, but we will not discuss these complications. *Routers* connect Steps together by examining the answer produced by the Step (technically, the answer produced by the Step's Worker) and deciding where to send the Material for subsequent processing. The LabFlow framework also includes classes for *Materials* and *States* which are analogous to those in LabBase, and a *LabFlow* class to represent entire workflows. LabFlow contains an "execution engine" that runs a workflow by invoking its Step repeatedly. A LabFlow can be packaged as a *subworkflow* and used subroutine-style within another workflow. A key limitation in our current software is that a given LabFlow operates on a single kind of Material.

The system uses States to keep track of "where" each Material is in a LabFlow. We have already mentioned that States are used to represent the input queues of each Step. In addition, each Step has three other built-in States. One of these holds Materials that are actively being processed by the Step. A second holds Materials for which the Step (or its Worker) discovered a problem that contraindicates further progress, e.g., the Step checking for *E. coli* contamination would place contaminated Materials in this State. The final built-in State holds Materials that were being processed by the Step when its Worker or its component crashed; sometimes this is just a coincidence, but often it reflects a bug in the Worker or component software. In addition to these built-in States, the workflow designer is responsible for defining States that represent successful completions of the workflow. The designer may also define additional failure States, and States that represent points where the protocol "pauses".

The biggest job in creating a LIMS is to develop or acquire the components since these do the actual computational work of the project. The next biggest job is to develop the Workers since these must accommodate the idiosyncrasies of the components. A typical system contains many Workers (dozens or more), and new ones must be developed as the computational needs of the project change. It is reasonable to expect that Workers may be reused in different applications, e.g., the sequence-assembly Worker in our cDNA-sequencing example might be reused in a genomic sequencing project. It is also reasonable to imagine that two or more different Workers might be implemented for the same job using different components, e.g., we might develop one sequence-assembly Worker based on phrap [19] and another based on TIGR Assembler [20]. It is also reasonable to expect that the same component might be used for several purposes, e.g., we might use a fast sequence alignment program, such as FASTA [21] or crossmatch [22], for both vector-stripping and contamination-checking. For these reasons, it makes sense to organize the collection of Workers as a class library with well-defined interfaces that are separate from their implementations, and to allow Workers to call each other. In the long run, success with our method (and probably any other modular approach to LIMS construction) depends on the accumulation of a well-designed library of good Workers.

After Workers are developed, what remains is to connect them together into a complete workflow. There are two main tasks: a Step must be created for each Worker, and Routers must be defined to connect the Steps together. The main work in defining a Step is to determine the mapping between the field-names used by the Worker and those used by the workflow as a whole. (These field-names may be different since Workers are written for reuse). Routers are generally straightforward for success-cases, but can be tricky for failure-cases; often, in the early days of a project, all failures are sent to a catch-all Worker that reports the event to laboratory supervisors.

Let us apply these ideas to our running example. We will model the database as suggested in the previous section, i.e., with two kinds of Materials, namely, sequence-reads and sequence-assemblies. Since a given LabFlow can operate on only one kind of Material, the overall system will need two LabFlows. We will only describe the first. The sequence-read LabFlow needs Workers for robot-control, base-calling, vector-clipping, quality-screening, and review by laboratory personnel.

Robot-control software generally comes with the machine and cannot be modified by the customer; often the software runs on a dedicated computer, and can only be operated by a person entering commands directly at that computer. The Worker, in such cases, helps the human operator coordinate the robot with the rest of the system. Assume for purposes of the example, that the outputs of the robotic procedure are (i) a collection of plated clones, and (ii) a collection of plated sequencing-templates derived from those clones, and that these plates are bar-coded. The most important coordination task is to record the bar-codes of the plates in such a way that the each clone-plate is associated with the corresponding template-plate (so that subsequent sequence data can be associated with the correct clone). The Worker software for doing this might be no more than a Web-based program that accepts bar-codes (entered by the operator using a bar-code wand) two-at-a-time, and passes each pair back to the Step.

Next comes base-calling. Assume that we use phred [23] for this purpose, and that we wish to run phred in real-time on the data stream generated by the sequencing machine. This requires that the Worker invoke phred on a computer directly attached to the sequencing instrument. The output of phred consists of a collection of files that can be stored in a network-accessible file directory. The Worker can monitor the contents of this directory and report back when the procedure completes.

The next tasks, vector stripping and quality screening, are purely computational and can be programmed straightforwardly to take sequence data as input, invoke the relevant program, and capture the outputs for further processing.

The final job is for laboratory personnel to review the results. The Worker for this Step might be a Web-based program with three stages. First, it presents a work-list showing sequence-reads ready for review and allows a human operator to select one or more sequences to review. Then, for each selected sequence, it presents an appropriate display of the results and allows the operator to indicate whether the sequence passes or fails. Finally, it passes the operator's decisions back to the Step. Based on this information, the Router for the Step would decide whether to move each sequence to a final success State or to a failure State

The example illustrates many of the kinds of Workers that arise in practice. Some Workers control or monitor external instruments. Some require human interaction. Some invoke external programs which may have to execute on a particular computer. As a general rule, we have limited ability (if any) to modify these external programs. Though not illustrated, some Workers interact with external servers, e.g., a BLAST server to analyze sequence-assemblies. Also not illustrated, some Workers are so simple that they may be implemented as Perl modules that run in the same process as the LabFlow engine itself.

LabFlow Details

The LabFlow execution engine runs a workflow by invoking its Steps. The execution engine groups Steps into "execution units" and initiates one or more operating-system processes to run each execution unit. In simple cases, the software allocates one process per Step. Each process sits in an infinite loop executing each of its Steps that has work to do, and sleeping when all Steps are quiescent. The execution engine is responsible for monitoring the status of these processes, starting new ones if any should fail, and shutting down the entire entourage on demand.

Figure 3 illustrates what happens when a Step is invoked. The Step use the LabBase get operation to discover whether there are any Materials in its input queue, and, if so, to obtain the fields of each Material needed to perform its computation. Usually, these fields were placed in the database by a previous Step and are obtained via the LabBase mechanism for accessing Step-data as if it were directly attached to Materials. The Step changes the field-names in the retrieved Material to the field-names used by its Worker, and passes the Material to its Worker as an ordinary Perl5 object.

The Worker converts the object to whatever format its component requires and passes it to the component. The component performs its computation and returns the results to the Worker. The Worker converts the component's result into a Perl5 object with the format required by LabFlow, and passes it back to the Step. The Step changes the field-names in the result to those used in the database. It associates the result with the Material, and forwards the ensemble to its Routers.

The Routers decide where the Material should be sent next. This is usually another Step (technically, the State representing the input queue of the next Step), but it may be a failure State, a terminal success State, or any other State. The Routers can examine the results of the Worker and any fields of the Material. The Material may be routed to multiple next States which allows the Material to flow down multiple paths in parallel.

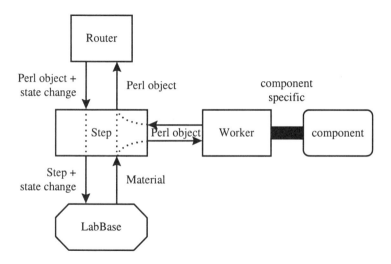

Figure 3: Execution of a Step.

The Step attaches the routing decision to the result, and then uses the LabBase put operation to store the result in the database as a LabBase-Step and to move the Material to its next State or States. The software performs the database update and routing tasks as an atomic transaction to ensure that even if the system were to crash at an inopportune moment, the database would not contain the results of the Step while the Material remained in the previous State, nor would the Material be moved to the next State without the result being placed in the database. LabFlow relies on LabBase mechanisms described previously to achieve the required atomicity.

The software can also operate on lists of Materials instead of just one Material at-a-time as described here.

290

LabFlow Summary

LabFlow makes it easier to create a LIMS by providing facilities for orchestrating the execution of components. Workers encapsulate the vagaries of invoking components, capturing their outputs, and coping with failures. Steps handle database access. Routers control the order in which components execute. States keep track of the progress of Materials as they advance through the workflow. The LabFlow engine acts as a scheduler to carry out the work specified by the above elements.

It would be useful to extend the system to handle workflows involving multiple Materials. Suppose we expand our example to include expression monitoring, in which cDNAs are arrayed on a chip and probed with different libraries. It is natural to think of this as a two dimensional process whose materials are chips (which we would like to think of as groups of cDNAs) and libraries, and whose result is a two-dimensional matrix indicating expression level as a function of cDNA and library. In broad terms, the workflow for this process would include the following Steps: (1) prepare chip for probing; (2) prepare library for probing; (3) probe; (4) analyze the results. It seems intuitive to regard the chip as the Material for step (1), the library as the Material for step (2), and the pair as the Materials for steps (3) and (4). This complicates our previously simple view of Materials "flowing through" a workflow. We may need to adopt two perspectives: a sample-tracking view, in which we regard a workflow as describing how a Material travels from Step-to-Step through a process, and a task-completion view, in which we regard a workflow as describing the process needed to accomplish an arbitrary task. While more general, the latter is also more complex, and we have yet to work out all its implications.

Another useful extension would be to handle workflows involving related Materials. We earlier mentioned three important kinds of inter-Material relationships, namely, derived-from, grouping, and part/whole. If LabBase and LabFlow were extended to represent these relationships in a direct manner, it would allow us to represent the example database and workflow more naturally. Instead of combining clones and sequence-reads into a single kind of Material, each could exist independently. And instead of coordinating multiple reads per clone in the sequence-assembly workflow, we could handle this through an explicit clone-analysis workflow. We see this as an important benefit, since the central theme of our work is to simplify LIMS construction by eliminating arcane design tasks and reducing what remains to its most natural form.

Conclusion

The LabBase and LabFlow systems described in this chapter are the latest in a series of systems we have built to tackle the problems of data management and workflow management for large scale biological research projects. We and our colleagues have used the predecessor systems for several projects at the Whitehead/MIT Center for

Genome Research and are beginning to use the current systems for projects there and elsewhere. Though the software is incomplete in many ways, it is proven technology that in our hands, at least, greatly reduces the work of creating laboratory informatics systems.

While we welcome others to use our software, we believe there is greater value in the ideas. The total quantity of code is modest, comprising about 10,000 lines of Perl5. It would not be hard to reproduce the ideas in other contexts, and we encourage others to do so.

References

1. Sargent, R., D. Fuhrman, T. Critchlow, T.D. Sera, R. Mecklenburg, G. Lindstrom, and P. Cartwright. *The Design and Implementation of a Database For Human Genome Research.* in Eighth International Conference on Scientific and Statistical Database Management. 1996. Stockholm, Sweden: IEEE Computer Society Press.
2. Kerlavage, A.R., M. Adams, J.C. Kelley, M. Dubnick, J. Powell, P. Shanmugam, J.C. Venter, and C. Fields. *Analysis and Management of Data from High Throughput Sequence Tag Projects.* in 26th Annual Hawaii International Conference on System Sciences. 1993: IEEE Computer Society Press.
3. Clark, S.P., G.A. Evans, and H.R. Garner, *Informatics and Automation Used in Physical Mapping of the Genome,* in *Biocomputing: Informatics and Genome Projects,* D. Smith, Editor. 1994, Academic Press: New York. p. 13-49.
4. Rozen, S., L.D. Stein, and N. Goodman, *LabBase: A Database to Manage Laboratory Data in a Large-Scale Genome-Mapping Project.* IEEE Transactions on Engineering in Medicine and Biology, 1995. **14**: p. 702-709.
5. Stein, L.D., S. Rozen, and N. Goodman. *Managing Laboratory Workflow With LabBase.* in 1994 Conference on Computers in Medicine (CompMed94). 1994.
6. Stein, L., A. Marquis, E. Dredge, M.P. Reeve, M. Daly, S. Rozen, and N. Goodman. *Splicing UNIX into a Genome Mapping Laboratory.* in USENIX Summer 1994 Technical Conference. 1994: USENIX.
7. Goodman, N., S. Rozen, and L.D. Stein. *Building a Laboratory Information System Around a C++-based Object-Oriented DBMS.* in 20th International Conference on Very Large Data Bases. 1994. Santiago de Chile, Chile: The Very Large Data Bases (VLDB) Endowment Inc.
8. Mohan, C., G. Alonso, R. Guenthoer, M. Kamath, and B. Reinwald. *An Overview of the Exotica Research Project on Workflow Management Systems.* in 6th International Workshop on High Performance Transaction Systems. 1995. Asilomar, CA.
9. Mohan, C., *Tutorial: State of the Art in Workflow Management System Research and Products,* http://www.almaden.ibm.com/cs/exotica/sigmod96.eps. 1996, IBM Almaden Research Center.
10. Hollingsworth, D., *The Workflow Reference Model,* http://www.aiai.ed.ac.uk:80/project/wfmc/. 1994, Workflow Management Coalition.

11. Fayad, M. and D.C. Schmidt, *Object-Oriented Application Frameworks.* Communications of the ACM, 1997. **40**(10): p. 32-28.

12. Durbin, R. and J.T. Mieg, *A C. elegans Database*, Documentation, code and data available from anonymous FTP servers at lirmm.lirmm.fr, cele.mrc-lmb.cam.ac.uk and ncbi.nlm.nih.gov. 1991

13. Chen, I.-M.A. and V.M. Markowitz, *An Overview of the Object-Protocol Model (OPM) and OPM Data Management Tools.* Information Systems, 1995. **20**(5): p. 393-418.

14. McHugh, J., S. Abiteboul, R. Goldman, D. Quass, and J. Widom, *Lore: A Database Management System for Semistructured Data.* SIGMOD Record, 1997. **26**(3): p. 54-66.

15. Buneman, P., S. Davidson, G. Hillebrand, and D. Suciu. *A Query Language and Optimization Techniques for Unstructured Data.* in ACM Conference on Management of Data (SIGMOD). 1996. Montreal Quebec.

16. Altschul, S.F., W. Gish, W. Miller, and D.J. Lipman, *Basic Local Alignment Search Tool.* Journal of Molecular Biology, 1990. **215**: p. 403-410.

17. Cattell, R., *Object Data Management Revised Edition: Object-Oriented and Extended Relational Database Systems.* 1994, Reading, MA: Addison-Wesley.

18. Orfali, R., D. Harkey, and J. Edwards, *The Essential Distributed Objects Survival Guide.* 1996, New York: John Wiley & Sons.

19. Green, P., *PHRAP Documentation*, http://www.mbt.washington.edu/phrap.docs/phrap.html. 1996, University of Washington.

20. Sutton, G., O. White, M.D. Adams, and A.R. Kerlavage, *TIGR Assembler: A New Tool for Assembling Large Shotgun Sequencing Projects.* Genome Science & Technology, 1995. **1**: p. 9-19.

21. Pearson, W.R., *Rapid and Sensitive Sequence Comparison with FASTP and FASTA.* Methods in Enzymology, 1990. **183**: p. 63-98.

22. Green, P., *SWAT, CROSS_MATCH Documentation*, http://www.mbt.washington.edu/phrap.docs/general.html. 1996, University of Washington.

23. Green, P., *PHRED Documentation*, http://www.mbt.washington.edu/phrap.docs/phred.html. 1996, University of Washington.

INDEX

INDEX